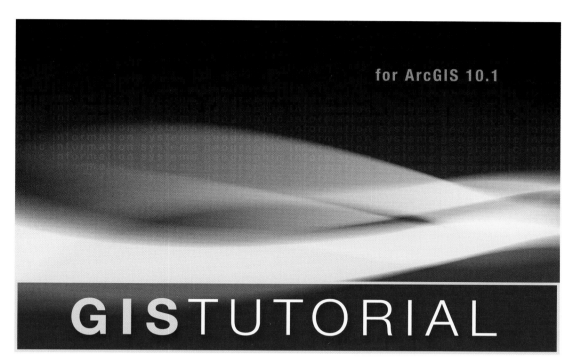

for ArcGIS 10.1

GISTUTORIAL

Basic Workbook

1

Wilpen L. Gorr
Kristen S. Kurland

Esri Press
REDLANDS|CALIFORNIA

Contents

Preface   ix

Acknowledgments   xi

Part I   *Using and making maps*

Chapter 1:   **Introduction   1**

Tutorial 1-1 Opening and saving a map document   2
Tutorial 1-2 Working with map layers   5
Tutorial 1-3 Navigating in a map document   13
Tutorial 1-4 Measuring distances   20
Tutorial 1-5 Working with feature attributes   23
Tutorial 1-6 Selecting features   27
Tutorial 1-7 Changing selection options   28
Tutorial 1-8 Working with attribute tables   33
Tutorial 1-9 Labeling features   39
   Assignment 1-1 Analyze population by race in the top 10 US states   42
   Assignment 1-2 Produce a crime map   44

Chapter 2:   **Map design   47**

Tutorial 2-1 Creating point and polygon maps using qualitative attributes   48
Tutorial 2-2 Creating point and polygon maps using quantitative attributes   58
Tutorial 2-3 Creating custom classes for a map   62
Tutorial 2-4 Creating custom colors for a map   66
Tutorial 2-5 Creating normalized and density maps   69
Tutorial 2-6 Creating dot density maps   74
Tutorial 2-7 Creating fishnet maps   76
Tutorial 2-8 Creating group layers and layer packages   83
   Assignment 2-1 Create a map showing schools in New York City by type   88
   Assignment 2-2 Create maps for military sites and congressional districts   90
   Assignment 2-3 Create maps for US veteran unemployment status   92

Chapter 3:   **GIS outputs   95**

Tutorial 3-1 Building an interactive GIS   96
Tutorial 3-2 Creating map layouts   103
Tutorial 3-3 Reusing a custom map layout   110
Tutorial 3-4 Creating a custom map template with two maps   112
Tutorial 3-5 Adding a report to a layout   118
Tutorial 3-6 Adding a graph to a layout   120
Tutorial 3-7 Building a map animation   122
Tutorial 3-8 Using ArcGIS Online   126
   Assignment 3-1 Create a dynamic map of historic buildings in downtown Pittsburgh   136
   Assignment 3-2 Create a layout comparing 2010 elderly and youth population compositions in Orange County, California   138
   Assignment 3-3 Create an animation for an auto theft crime time series   140
   Assignment 3-4 Create a shared map on ArcGIS Online   142

Part II Working with spatial data

Chapter 4:   **File geodatabases**   **145**
   Tutorial 4-1 Building a file geodatabase   146
   Tutorial 4-2 Using ArcCatalog utilities   149
   Tutorial 4-3 Modifying an attribute table   152
   Tutorial 4-4 Joining tables   155
   Tutorial 4-5 Creating centroid coordinates in a table   157
   Tutorial 4-6 Aggregating data   161
    Assignment 4-1 Investigate educational attainment   166
    Assignment 4-2 Compare serious crime with poverty in Pittsburgh   168

Chapter 5:   **Spatial data**   **171**
   Tutorial 5-1 Examining metadata   172
   Tutorial 5-2 Working with world map projections   175
   Tutorial 5-3 Working with US map projections   178
   Tutorial 5-4 Working with rectangular coordinate systems   180
   Tutorial 5-5 Learning about vector data formats   185
   Tutorial 5-6 Downloading US Census Bureau boundary maps   191
   Tutorial 5-7 Downloading and processing Census SF 1 data tables   193
   Tutorial 5-8 Downloading and processing American Community Survey (ACS) Census data   201
   Tutorial 5-9 Exploring raster basemaps from Esri web services   204
   Tutorial 5-10 Downloading raster maps from the USGS   206
   Tutorial 5-11 Exploring sources of GIS data from government websites   210
    Assignment 5-1 Compare heating fuel types by US counties   214
    Assignment 5-2 Create a map of Maricopa County, Arizona, voting districts, schools, and voting-age population using downloaded data and a web service image   216

Chapter 6:   **Geoprocessing**   **219**
   Tutorial 6-1 Extracting features for a study area   220
   Tutorial 6-2 Clipping features   224
   Tutorial 6-3 Dissolving features   226
   Tutorial 6-4 Merging features   229
   Tutorial 6-5 Intersecting layers   231
   Tutorial 6-6 Unioning layers   234
   Tutorial 6-7 Automating geoprocessing with ModelBuilder   239
    Assignment 6-1 Build a study region for Colorado counties   248
    Assignment 6-2 Dissolve property parcels to create a zoning map   250
    Assignment 6-3 Build a model to create a fishnet map layer for a study area   252

Chapter 7:   **Digitizing**   **255**

Tutorial 7-1 Digitizing polygon features   256
Tutorial 7-2 Digitizing line features   264
Tutorial 7-3 Digitizing point features   268
Tutorial 7-4 Using advanced editing tools   271
Tutorial 7-5 Spatially adjusting features   276
   Assignment 7-1 Digitize police beats   280
   Assignment 7-2 Use GIS to track campus information   282

Chapter 8:   **Geocoding**   **285**

Tutorial 8-1 Geocoding data by ZIP Code   286
Tutorial 8-2 Geocoding data by street address   292
Tutorial 8-3 Correcting source addresses using interactive rematch   297
Tutorial 8-4 Correcting street reference layer addresses   299
Tutorial 8-5 Using an alias table   304
   Assignment 8-1 Geocode household hazardous waste participants to ZIP Codes   306
   Assignment 8-2 Geocode immigrant-run businesses to Pittsburgh streets   308
   Assignment 8-3 Examine match option parameters for geocoding   310

Part III Analyzing spatial data

Chapter 9:   **Spatial analysis**   **313**

Tutorial 9-1 Buffering points for proximity analysis   315
Tutorial 9-2 Conducting a site suitability analysis   320
Tutorial 9-3 Using multiple ring buffers for calibrating a gravity model   323
Tutorial 9-4 Using data mining with cluster analysis   329
   Assignment 9-1 Analyze population in California cities at risk for earthquakes   334
   Assignment 9-2 Analyze visits to the Jack Stack public pool in Pittsburgh   336
   Assignment 9-3 Use data mining with crime data   338

Chapter 10:   ArcGIS 3D Analyst for Desktop   341

    Tutorial 10-1 Creating a 3D scene   342

    Tutorial 10-2 Creating a TIN (triangulated irregular network) from contours   344

    Tutorial 10-3 Draping features onto a TIN   348

    Tutorial 10-4 Navigating scenes   353

    Tutorial 10-5 Creating an animation   357

    Tutorial 10-6 Using 3D effects   359

    Tutorial 10-7 Using 3D symbols   362

    Tutorial 10-8 Editing 3D objects   365

    Tutorial 10-9 Using ArcGIS 3D Analyst for landform analysis   368

    Tutorial 10-10 Exploring ArcGlobe   372

    Assignment 10-1 Develop a 3D presentation for downtown historic sites   376

    Assignment 10-2 Topographic site analysis   378

    Assignment 10-3 3D animation of conservatory study area   380

Chapter 11:   ArcGIS Spatial Analyst for Desktop   383

    Tutorial 11-1 Processing raster map layers   384

    Tutorial 11-2 Creating a hillshade raster layer   389

    Tutorial 11-3 Making a kernel density map   391

    Tutorial 11-4 Extracting raster value points   396

    Tutorial 11-5 Conducting a raster-based site suitability study   399

    Tutorial 11-6 Using ModelBuilder for a risk index   405

    Assignment 11-1 Create a mask and hillshade for suburbs   416

    Assignment 11-2 Estimate heart-attack fatalities outside of hospitals by gender   418

Appendix A   Task index   421

Appendix B   Data source credits   425

Appendix C   Data license agreement   439

Appendix D   Installing the data and software   443

Preface

GIS Tutorial 1: Basic Workbook is the direct result of the authors' experiences teaching GIS to undergraduate and graduate students in several departments and disciplines at Carnegie Mellon University, high school students in a summer program at Carnegie Mellon University, as well as professionals in workshops.

GIS Tutorial 1 is a hands-on workbook with step-by-step exercises that take the reader from the basics of using ArcGIS for Desktop through performing many kinds of spatial analyses. Instructors can use this book for the lab portion of a GIS course, or individuals can use it for self-study. You can learn a lot about GIS concepts and principles by "doing" and we provide many short notes on a "just-in-time" basis to help this kind of learning.

The book has three parts. Part I, "Using and making maps," is essential for all beginning students. The chapters of part II, "Working with spatial data," have students download spatial data from government sources on the Internet and prepare the data for use in GIS. Part III, "Analyzing spatial data," puts GIS to work for more than just mapping, introducing many analytical tools that are unique to GIS. The parts of the book are largely independent of each other, and you can use them in the order that best fits your needs or your class's needs. Moreover, while it's best to cover the chapters of part I in order, you can use the chapters of parts II and III in any order.

Part I has three chapters. In chapter 1, readers learn the basics of working with existing GIS data and maps. Chapter 2 has the reader build many of the kinds of maps commonly used in practice. Chapter 3 covers the forms in which maps are used, including static map layouts, interactive maps, map animations, and maps published on the Internet.

With its five chapters, Part II is the longest and provides much expertise on building a GIS. The exercises in chapter 4 teach readers how to create geodatabases and import data into them, join tables, and carry out other data processing steps. Chapter 5 explores the basic spatial data formats used in GIS, shows readers how to download basemaps and related data from major governmental spatial data suppliers, and then how to prepare and import downloaded spatial data into ArcGIS. Chapter 6 covers geoprocessing tools and workflows needed to transform basemaps into study areas in a GIS. Included is an exercise on creating macros for automating multiple-step workflows. Editing spatial data is a vital part of GIS work, so chapter 7 teaches how to digitize and edit vector data and transform data to match geographic coordinates. In chapter 8, students learn how to transform textual address data into mapped points through the geocoding process.

Part III has three chapters on analyzing spatial data. Chapter 9 focuses on standard tools for spatial analysis, including buffers for proximity analysis, spatial joins and intersections for selecting areas that meet multiple criteria, and cluster analysis for data mining of attribute and spatial data. Chapters 10 and 11 introduce two ArcGIS extensions. Chapter 10 introduces ArcGIS 3D Analyst for Desktop, enabling students to create 3D scenes, conduct fly-through animations, and conduct landform analyses. Finally, chapter 11 introduces ArcGIS Spatial Analyst for Desktop for creating and analyzing raster maps, including density maps, site suitability surfaces, and risk index surfaces.

To reinforce the skills learned in the step-by-step exercises and to provoke critical problem-solving skills, short Your Turn assignments are included throughout each chapter with challenging homework assignments at the end of each chapter. The quickest way to increase GIS skills is to follow up step-by-step instructions with independent work, and the assignments provide these important learning components. Note that you must complete most Your Turn assignments to work on subsequent exercise steps.

This book comes with a DVD containing exercise and assignment data, completed exercises in the EsriPress\GIST1\MyExercises\FinishedExercises folder, and an evaluation code that can be used to download a 180-day version of ArcGIS 10.1 for Desktop Advanced. You need to install the software and data to perform the exercises and assignments in this book (see appendix D). For instructor resources and updates related to this book, go to esri.com/esripress.

Acknowledgments

We would like to thank all who made this book possible.

We have taught GIS courses at Carnegie Mellon University since the late 1980s, always with lab materials we had written. With the feedback and encouragement of students, teaching assistants, and colleagues, we eventually wrote what became this book. We are forever grateful for that encouragement and feedback.

Faculty members of other universities who have taught GIS using *GIS Tutorial 1: Basic Workbook* have also provided valuable feedback. They include Luke Ward of Rocky Mountain College; Irene Rubinstein of Seneca College; An Lewis of the University of Pittsburgh; George Tita of the University of California, Irvine; Walter Witschey of Longwood University; and Jerry Bartz of Brookhaven College.

We are very grateful to the many public servants and vendors who have generously supplied us with interesting GIS applications and data, including Kevin Ford of Facilities Management Services, Carnegie Mellon University; Barb Kviz of the Green Practices program, Carnegie Mellon University; Mike Homa of the Department of City Planning, City of Pittsburgh; staff members of the New York City Department of City Planning; Michael Radley of the Pittsburgh Citiparks Department; Pat Clark and Traci Jackson of Jackson Clark Partners; staff of the Pennsylvania Resources Council; Chief Nathan Harper of the Pittsburgh Bureau of Police; former Mayor Robert Duffy of the City of Rochester, New York; Kirk Brethauer of Southwestern Pennsylvania Commission; and employees of several spatial data vendors (including InfoUSA; i-cubed; Information Integration & Imaging, LLC; TomTom MultiNet; and Tele Atlas).

Finally, we are much indebted to the wonderful staff at Esri Press for editorial expertise and beautiful design work, and efficient production and distribution of our book. It's a pleasure working with these dedicated and talented professionals.

Part I
Using and making maps

1

Introduction

This first chapter familiarizes you with some of the basics of the ArcGIS for Desktop software as well as the general field of GIS. You work with map layers and underlying attribute data tables for US states, cities, counties, and streets. These layers are made up of spatial vector features consisting of points, lines, and polygons. Each geographic feature has a corresponding data record, and you work with both features and their data records. Vector map layers and raster images are increasingly found on Esri's website, ArcGIS.com, and on other cloud websites. You will learn more about ArcGIS Online in chapter 3 and about other sources for GIS layers in chapter 5.

Learning objectives

- *Open and save a map document*
- *Work with map layers*
- *Navigate in map documents*
- *Measure distances*
- *Work with feature attributes*

- *Select features*
- *Change selection properties*
- *Work with attribute tables*
- *Label features*

Tutorial 1-1

Opening and saving a map document

ArcMap is the primary mapping component of ArcGIS for Desktop software from Esri. Esri offers three licensing levels of ArcGIS for Desktop, each with increasing capabilities: Basic, Standard, and Advanced. Together, ArcMap, ArcCatalog, ArcScene, and ArcGlobe—all of which you will use in this book—make up ArcGIS for Desktop, the world's most popular GIS software.

Launch ArcMap

1 On the Windows taskbar, click Start, then All Programs > ArcGIS > ArcMap 10.1. Depending on your operating system and how ArcGIS and ArcMap have been installed, you may have a different navigation menu.

2 In the resulting ArcMap - Getting Started window, click Existing Maps > Browse for more.

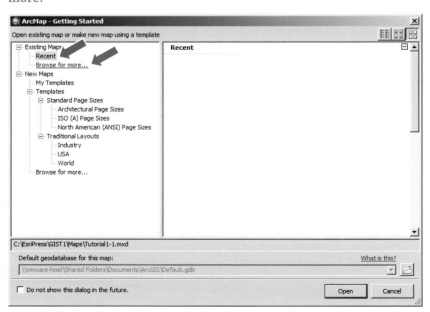

Open an existing map document

1-1
1-2
1-3
1-4
1-5
1-6
1-7
1-8
1-9
A1-1
A1-2

1 In the Open ArcMap Document dialog box, browse to the drive where you installed the \EsriPress\GIST1\Maps\ folder (for example, C:\EsriPress\GIST1\Maps\), select Tutorial1-1.mxd, and click Open.

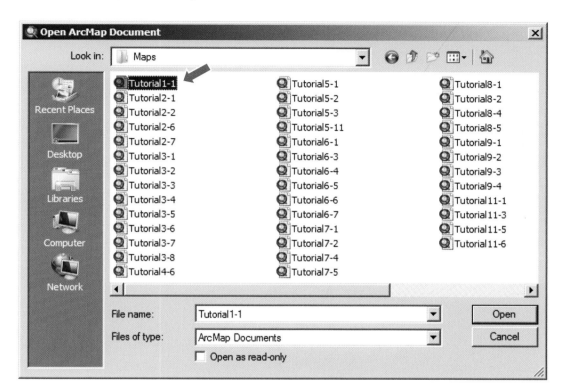

2 Click the Tutorial1-1.mxd (or Tutorial1-1) icon.

3 Click Open. The Tutorial1-1.mxd map document opens in ArcMap (see image on the next page) showing a map consisting of the US States layer (zoomed to the 48 contiguous states). The US Cities layer (not yet turned on) is the subset of cities with population greater than 300,000. The left panel of the ArcMap window is the table of contents. It serves as a legend for the map plus has several other uses you will learn about. Note that the Tools toolbar, which is floating on the bottom of the screen on the next figure, may be docked somewhere in the interface. If you wish, you can anchor it by clicking in its top area, dragging it to a side or top of the map display window, and releasing when you see a thin rectangle materialize. If you do not see the Tools toolbar at all, click Customize > Toolbars > Tools to make it visible. You will learn to use many of the tools in this toolbar throughout this book.

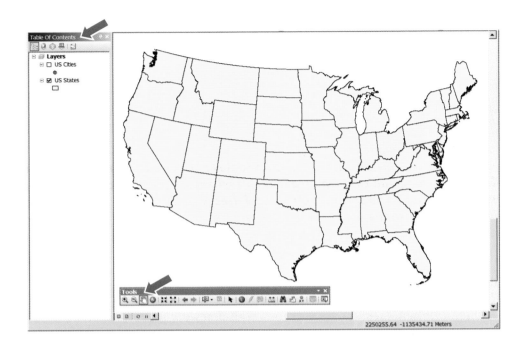

Save the map document to a new location

You will save all files that you modify or create while working through the tutorials in this book in the \EsriPress\GIST1\MyExercises chapter folders.

1 On the Menu bar, click File > Save As.

2 Save your map document as **Tutorial1-1.mxd** to the Chapter1 folder of MyExercises.

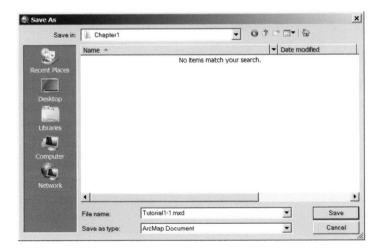

Tutorial 1-2

Working with map layers

Map layers are references to vector data sources such as points, lines, and polygons, raster images, and so forth representing spatial features that can be displayed on a map. ArcMap displays map layers from a map document such as Tutorial1-1.mxd, but the map document does not contain copies of the map layers. The map layer files remain external to the map document wherever they exist on computer storage media, whether on your computer, a local area network, or on an Internet server. Next, you will use the map document's table of contents to display the map layers in the document.

Turn a layer on and off

Before GIS existed, mapmakers drew separate layers on clear plastic sheets and then carefully stacked the sheets to make a map composition. Now with GIS, working with layers is much easier.

1 Save your map document as **Tutorial1-2.mxd** to the Chapter1 folder.

2 In the table of contents, select the small check box to the left of the US Cities layer. A check mark appears if the layer is turned on. Note that if the table of contents accidentally closes, you can click Windows > Table of Contents to reopen it.

3 In the table of contents, clear the check box to the left of the US Cities layer to turn the layer off.

Add and remove map layers

You can add map layers to the table of contents from their storage locations.

1 Click the Add Data button ✛.

2 In the Add Data dialog box, click the Connect to Folder button 🗁⁺.

3 In the Connect to Folder dialog box, browse to Computer, click the drive where you installed the GIST 1 data (for example, C:\), and click OK. You need to browse to the root drive only and not go beyond this. Once this connection is made, you can easily browse to any folder on this computer drive. If your data is stored on an external drive such as a jump drive, you will need to connect to that drive (for example, E:\).

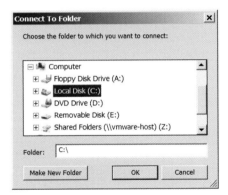

4 In the Add Data dialog box, browse to \EsriPress\GIST1\Data\UnitedStates.gdb, click the COCounties layer > Add. ArcMap randomly picks a color for the Colorado counties layer. You will learn how to change the color and other layer symbols later. You may receive a warning about the map's geographic coordinate system.

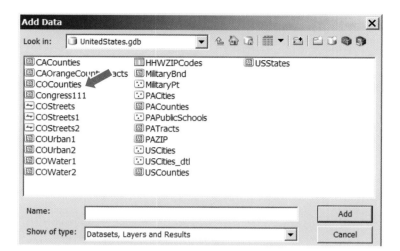

1-1

1-2

1-3

1-4

1-5

1-6

1-7

1-8

1-9

A1-1

A1-2

5 Close the Geographic Coordinate Systems Warning window. ArcMap will place the new layer with Colorado counties correctly over the state of Colorado because all map layers have coordinates tied to the earth's surface.

6 In the table of contents, right-click COCounties and click Remove. This action removes the map layer from the map document but does not delete it from its original storage location, UnitedStates.gdb.

Use relative paths

When you add a layer to a map, ArcMap stores its path in the map document. When you open a map, ArcMap locates the layer data it needs using these stored paths. If ArcMap cannot find the data for a layer, the layer will still appear in the ArcMap table of contents, but of course it will not appear on the map. Instead, ArcMap places a red exclamation mark (!) next to the layer name to indicate that its path needs repair. You can view information about the data source for a layer and repair it by clicking the Source tab in the Layers Properties window.

Paths can be absolute or relative. An example of an absolute path is C:\EsriPress\GIST1\ Data\UnitedStates.gdb\USCities. To share map documents saved with absolute paths, everyone who uses the map document must have exactly the same paths to map layers on his or her computer. Instead, the relative path option is favored.

Relative paths in a map specify the location of the layers relative to the current location of the map document on disk (.mxd file). Because relative paths do not contain drive letter names, they enable the map and its associated data to point to the same directory structure regardless of the drive or folder in which the map resides. If a map document and associated folders are moved to a new drive, ArcMap will still be able to find the maps and their data by traversing the relative paths.

1 On the Menu bar, click File > Map Document Properties. Notice the option is set to "Store relative pathnames to data sources." This is for the current map document only.

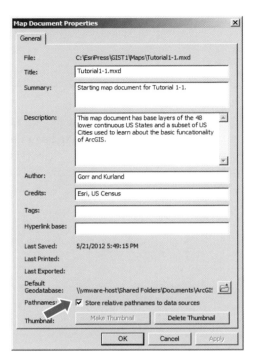

2 Click OK.

Next, you set an option to have all future map documents use relative paths.

3 On the Menu bar, click Customize > ArcMap Options > General tab.

4 In the ArcMap Options dialog box, General section, select the check box beside "Make relative paths the default for new map documents" > OK.

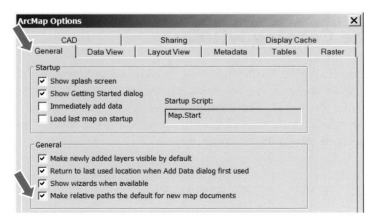

Drag and drop a layer from the Catalog window

The Catalog window, with its many utility functions, allows you to explore, maintain, and use GIS data. From Catalog, you drag and drop a map layer into the table of contents as an alternative method of adding data.

1 On the Menu Bar, click Windows > Catalog.

2 In the Catalog window, navigate to UnitedStates.gdb in the Data folder.

3 Drag COCounties into the top of the Table Of Contents window. The map layers in the table of contents draw in order from the bottom up, so if you dropped COCounties below US States, US States would cover COCounties. If COCounties is covered, remove it and drag and drop it again from Catalog, this time above US States.

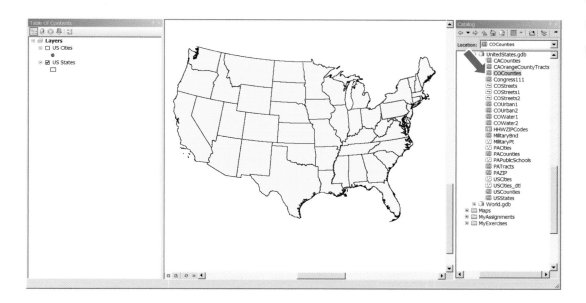

1-1
1-2
1-3
1-4
1-5
1-6
1-7
1-8
1-9
A1-1
A1-2

Use Auto Hide for the Catalog window

Notice that when you opened the Catalog window, it opened in pinned-open mode, which keeps the window open and handy for use, but covers part of your map. The Auto Hide feature of this application window and of other application windows (such as Table of Contents and Search) keeps the windows available for immediate use, but hides them in between uses so that you have more room for your map.

1 Click the Auto Hide button [image] on top of the Catalog window. The window closes but leaves a Catalog tab on the right side of the ArcMap window.

2   Click the Catalog button [Catalog] and the Catalog window opens.

Next, you simulate having completed a Catalog task by clicking the map document. The window auto hides.

3   Click any place on the map or table of contents.

You can pin the window open again, which you do next.

4   Click the Catalog button and click the Unpinned Auto Hide button ⊞. That pins the Catalog window open until you click the pin again to auto hide or close the window. Try clicking the map or table of contents to see that the Catalog window remains open.

5   Close the Catalog window.

> ### *YOUR TURN*
>
> Use the Add Data or Catalog button to add COStreets, also found in UnitedStates.gdb. These are street centerlines for Jefferson County, Colorado. You may have difficulty seeing the streets because they occupy only a small area of the map (look carefully above the center of Colorado). Later, in tutorial 1-3, you learn how to zoom in for a closer look.

Change a layer's display order

Next, you change the drawing order of layers. For this you must have the List By Drawing Order button selected.

1   In the table of contents, click the List By Drawing Order button ⁑ and turn on the US Cities layer.

2   Drag the US Cities layer to the bottom of the Table of Contents. Because ArcMap draws the US Cities layer first now, the US States and Counties layers cover its point markers.

3 Drag the US Cities layer to the top of the Table of Contents and drop it. ArcMap now draws the US Cities last, so you can see its points again.

Change a layer's color

One of the nicest capabilities of ArcGIS is how easy it is to change colors and other symbols of layers. First you change the color fill of a layer's polygons.

1 In the table of contents, click the COCounties layer's legend symbol. The legend symbol is the rectangle below the layer name in the table of contents.

2 In the resulting Symbol Selector dialog box, click the Fill Color button [ ].

3 In the color palette, click the Tarragon Green tile (column 6, row 5).

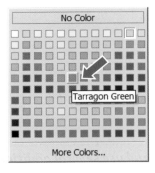

4 Click OK.

Change a layer's outline color

1 In the table of contents, click the layer's legend symbol for COCounties.

2 In the Symbol Selector dialog box, click the Outline Color button.

3 In the color palette, click the Black tile and click OK.

YOUR TURN

Change the color of the COStreets layer, choosing a light shade of gray (20%). You will see the results later. Save your map document.

1-1
1-2
1-3
1-4
1-5
1-6
1-7
1-8
1-9
A1-1
A1-2

Tutorial 1-3

Navigating in a map document

If you are zoomed out as far as possible in a map document, seeing the entire map, this is known as full extent. You can zoom in to any area of the map resulting in that area filling the map window, giving you a close-up view. The current view of the map is its current extent. You can zoom out, pan, and use several additional means of moving about in your map document including the Magnifier window for close-up views without zooming in, the Overview window that shows where you are on the full map when zoomed in, and spatial bookmarks for saving a map extent for future use.

Zoom In

1 Save your map document as **Tutorial1-3.mxd** to the Chapter1 folder.

2 On the Tools toolbar, click the Zoom In button 🔍 .

3 Press and hold down the mouse button on a point above and to the left of the state of Florida.

4 Drag the pointer to the bottom and to the right of the state of Florida and release the mouse button. The process you performed in steps 2 and 3 is called "dragging a rectangle."

Fixed Zoom In and Zoom Out

This is an alternative for zooming in by fixed amounts.

1 On the Tools toolbar, click the Fixed Zoom In button. This zooms in a fixed distance on the center of the current display.

2 Click the map to zoom in centered on the point you pick.

3 On the Tools toolbar, click the Fixed Zoom Out button. This zooms out a fixed distance from the center of the current zoomed display.

Pan

Panning shifts the current display in any direction without changing the current scale.

1 On the Tools toolbar, click the Pan button.

2 Move the pointer anywhere into the map view.

3 Pressing the left mouse button, drag the pointer in any direction.

4 Release the mouse button.

Full, previous, and next extent

The following steps introduce tools that navigate through views you have already created:

1 On the Tools toolbar, click the Full Extent button. This zooms to a full display of all layers, regardless of whether they are turned on or turned off. Notice that this zooms to all US States and territories.

2 On the Tools toolbar, click the Go Back to Previous Extent button. This returns the map display to its previous extent.

3 Continue to click this button to step back through all of the views.

4 On the Tools toolbar, click the Go to Next Extent button. This moves forward through the sequence of zoomed extents you have viewed.

5 Continue to click this button until you reach the contiguous 48 states.

1-1
1-2
1-3
1-4
1-5
1-6
1-7
1-8
1-9
A1-1
A1-2

> ***YOUR TURN***
>
> On the Menu bar, click Customize > ArcMap Options and explore how to control zoom and pan options with your mouse wheel.

Zoom to layer

Map layers have their own extents and you can zoom to each layer's extent.

1 In the table of contents, right-click the COCounties layer and click Zoom To Layer. This zooms to the extent of the counties in Colorado.

> ***YOUR TURN***
>
> Zoom to the extent of the layer COStreets (streets in Jefferson County, Colorado). Leave your map zoomed in to the streets.

Open the Magnifier window

The Magnifier window works like a magnifying glass. As you pass the window over the map display, you see a magnified view of the location under the window.

1 On the Menu bar, click Windows > Magnifier.

2 Drag the Magnifier over an area of the map to see a crosshair for area selection, and then release to see the zoomed details.

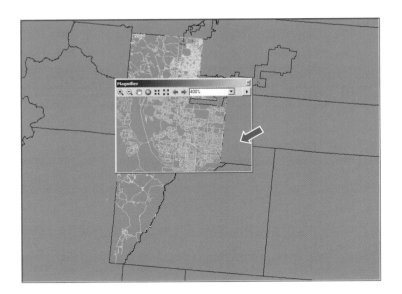

3 Drag the Magnifier window to a new area to see another detail on the map.

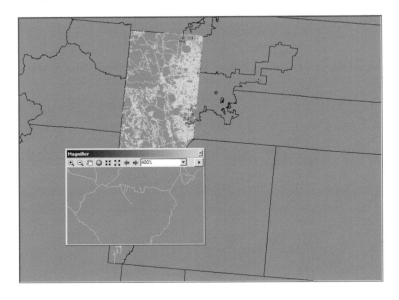

Change the Magnifier properties

1 Right-click the title bar of the Magnifier window and click Properties.

2 Change the Magnify By percentage to 800% if it is not already at that power, and click OK.

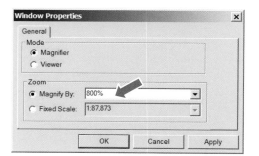

3 Drag the Magnifier window to a different location and see the resulting view.

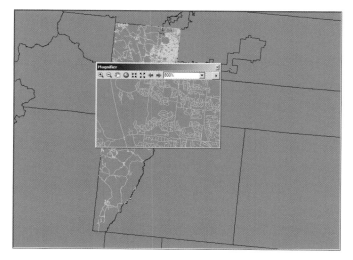

4 Close the Magnifier window.

Use the Overview window

The Overview window shows the full extent of the layers in a map. A box in the Overview window shows the currently zoomed area. You can move the box to pan the map display. You can also make the box smaller or larger to zoom the map display in or out.

1 Zoom to the previous extent until you see the 48 contiguous states.

2 Zoom to a small area of the map in the northwest corner of the US (two or three complete states).

3 On the Menu bar, click Windows > Overview. The Overview window shows the current extent of the map highlighted with a red rectangle.

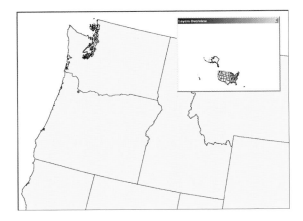

4 Move the cursor to the center of the red box and drag to move it to the northeast corner of the US. The extent of the map display updates to reflect the changes made in the Layers Overview window.

Note that if you right-click the top of the Layers Overview window and click Properties, you can modify the display, including the reference layer for map extents.

5 Close the Layers Overview window.

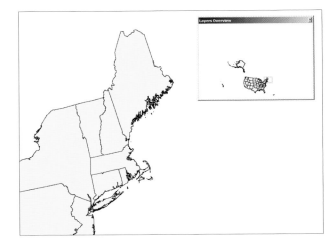

YOUR TURN

Practice using the Overview window using the Colorado Counties and Colorado Streets layers. In the Overview Properties window, set the Reference layer to COCounties. Close the Layers Overview window.

Use spatial bookmarks

Spatial bookmarks save the extent of a map display so you can return to it later without having to use Pan and Zoom tools.

1 In the table of contents, turn off the layers COCounties and COStreets.

First, use a bookmark that was already created.

2 On the Menu bar, click Bookmarks > 48 Contiguous States.

3 Zoom to the state of Florida.

Next, you create a new bookmark.

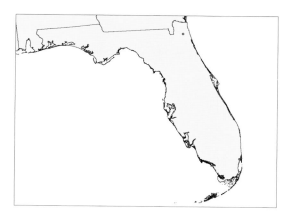

4 Click Bookmarks > Create Bookmark, and type **Florida** in the Bookmark Name field.

5 Click OK.

6 Click Bookmarks > 48 Contiguous States.

7 Click Bookmarks > Florida.

YOUR TURN

Create spatial bookmarks for the states of California, New York, and Texas. Try out your bookmarks. Save your map document.

1-1
1-2
1-3
1-4
1-5
1-6
1-7
1-8
1-9
A1-1
A1-2

Tutorial 1-4

Measuring distances

Maps have coordinates enabling you to measure distances along paths that you choose with your mouse and cursor.

Change measurement units

While a map's coordinates are in specific units such as feet or meters, you can set the measurement tool to gauge distances in any relevant units.

1 Save your map document as **Tutorial1-4.mxd** to the Chapter1 folder.

2 Click Bookmarks > 48 Contiguous States, then zoom to the state of Washington. Washington is the uppermost western state of the 48 contiguous states.

3 On the Tools toolbar, click the Measure button ▦.

The Measure window opens with the Measure Line tool enabled. The current map units are meters, but miles are more familiar in the United States, so change the units to Miles.

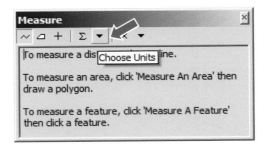

4 In the Measure window, click the Units button.

5 Click Distance > Miles, and leave the Measure window open.

Measure the width of Washington State

1 Click the westernmost boundary of the state of Washington.

You do not need to match the exact selections made in the following image. Any
measurement will demonstrate the procedure.

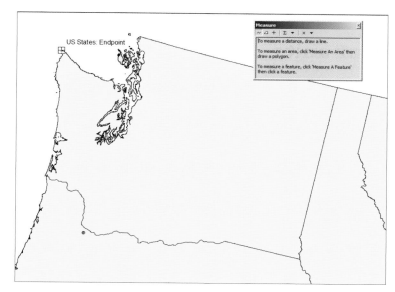

2 Move the mouse in a straight line to the eastern boundary of Washington until you reach
its eastern edge, then double-click the edge. The distance should be around 335 miles.

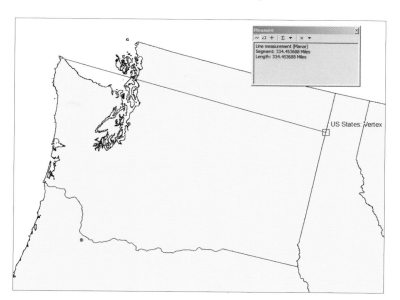

3 Close the Measure window.

YOUR TURN

Zoom to the 48 contiguous states and measure the north–south distance of the continental United States from the southern tip of Texas to the northern edge of North Dakota. This distance is approximately 1,600 miles. Measure the east–west distance of the continental United States from Washington to Maine. This distance should be approximately 2,500 miles. Zoom to and measure the north-south distance (top to bottom) of the state of Washington. The distance should be roughly 250 miles. Close the Measure window when finished. Save your map document.

Tutorial 1-5

Working with feature attributes

Graphic features of map layers and their data records are connected, so you can start with a feature and view its record. You can also find features on a map using feature attributes.

Use the Identify tool

To display the data attributes of a map feature, you can click the feature with the Identify tool. This tool is the easiest way to learn something about a location on a map.

1 Save your map document as **Tutorial1-5.mxd** to the Chapter1 folder.

2 Click Bookmarks > 48 Contiguous States.

3 On the Tools toolbar, click the Identify button ⓘ. Click anywhere on the map.

4 In the Identify window, click the Identify from list > US States.

5 Click inside the state of Texas. The state temporarily flashes and its attributes appear in the Identify dialog box. Note one of the field values for the state, such as Hispanic population, from the 2010 US Census.

Next, you use the Identify tool's options to control which features it processes.

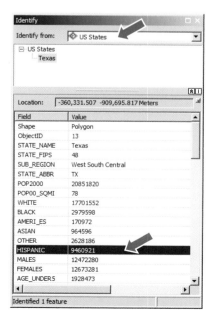

6   In the Identify window, click the Identify from drop-down list > US Cities.

7   Click the red circular point marker for Houston (at the southeastern side of Texas). Make sure the point of the arrow is inside the circle when you click the mouse button. Notice which feature flashes—that is the feature for which you get information. This table includes only basic population data for 2000 and 2007.

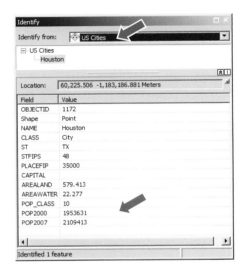

8   Continue clicking a few other cities to see the identify results. Hold down the SHIFT key to retain information for more than one city. Then click the name of a city in the top panel of the Identify window to view that city's information.

Use advanced Identify tool capabilities

You can use the Identify tool to navigate and create spatial bookmarks.

1   Without holding down the SHIFT key, click Houston with the Identify tool.

2   Right-click the name Houston in the Identify window and click Flash. This flashes Houston's point marker.

3   Right-click the name Houston again and click Zoom To. The map display zooms to Houston, Texas. ArcMap identifies the US Cities only because its layer is set in the dialog box.

4 Right-click the name Houston once again and click Create Bookmark. A bookmark of this zoomed area, called Houston, will automatically be created.

5 Close the Identify window.

6 Click Bookmarks > 48 Contiguous States.

7 Click Bookmarks > Houston.

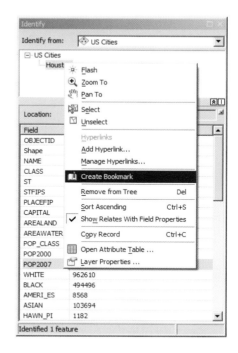

YOUR TURN

Turn on layers for COCounties and COStreets. Restrict the Identify results to the COCounties layer and identify Colorado counties. Practice making bookmarks for various Colorado counties using the Identify tool. Close the Identify window.

Find features

Use the Find tool to locate features in a layer or layers based on their attribute values. You can also use this tool to select, flash, zoom, bookmark, identify, or unselect the feature in question.

1 Click Bookmarks > 48 Contiguous States.

2 On the Tools toolbar, click the Find button 🔍.

3 Click the Features tab.

4 Type **Boston** as the feature and click Find. The results appear in the bottom section of the Find window.

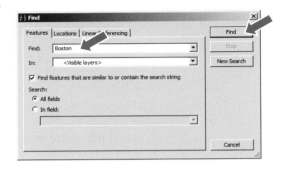

5 Right-click the city name Boston.

6 Click Zoom To. The extent zooms to the city of Boston, Massachusetts.

YOUR TURN

Find other US cities and practice showing them using other find options such as Flash Features, Identify Feature(s), and Set Bookmark. When finished, clear any selected features, and zoom to the 48 Contiguous States bookmark. Save your map document.

Tutorial 1-6

Selecting features

You can work with a subset of one or more features in a map layer by selecting them. For example, before you move, delete, or copy a feature (as you will learn about in future chapters), you must select it. Selected features appear highlighted in the layer's attribute table and on the map.

Use the Select Features tool

1   Save your map document as **Tutorial1-6.mxd** to the Chapter1 folder.

2   Click Bookmarks > 48 Contiguous States.

3   Turn off the COStreets and COCounties layers.

4   On the Tools toolbar, click the Select Features button ⟋ ▾.

5   Click inside Texas.
This action selects Texas and
highlights it with a blue outline.

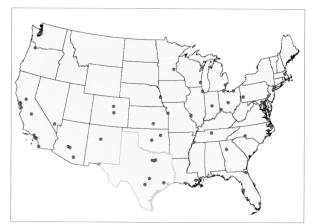

6   Hold down the SHIFT key and
click inside the four states adjacent
to Texas.

7   Save your map document.

Tutorial 1-7

Changing selection options

Sometimes you will want to produce a map with certain features selected. Then it is desirable to be able to change the selection color for the purpose at hand.

1 Save your map document as **Tutorial1-7.mxd** to the Chapter1 folder.

2 On the Menu bar, click Selection > Selection Options.

3 In the Selection Tools Settings frame, click the color box.

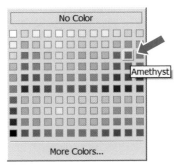

4 Click Amethyst (second column from the right, row 3) as the new selection color and click OK.

5 On the Menu bar, click Selection > Zoom to Selected Features. The map will zoom to the selected features and the selection color for map features will now be amethyst.

6 On the Menu bar, click Selection > Clear Selected Features.

Change selection properties by layer

In addition to changing the color of selected features, you can change the selection symbol for the entire map or for individual layers.

1 In the table of contents, right-click US Cities, and click Properties. The resulting Layer Properties window is one that you will use often. It allows you to modify many properties of a map layer.

2 Click the Selection tab and Symbol button.

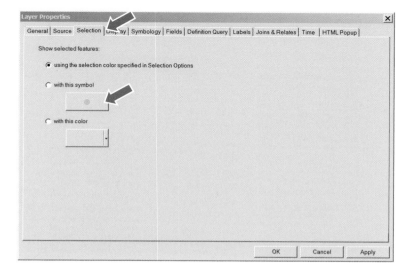

3 Click Square 2, Amethyst color, and size 12, and click OK > OK.

4 Click a city feature to see the new selection symbol, color, and size.

5 Clear the selected features.

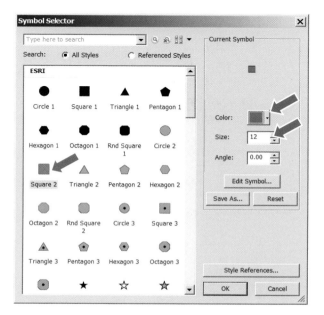

YOUR TURN

Turn on and zoom to layers COCounties and COStreets. Change the selection for Colorado counties (COCounties) to an Amethyst fill color with no outline. Select some counties to test the selection properties. When finished, turn off the COCounties layer. Change the selection color of the COStreets layer to Amethyst and select some streets.

Set selectable layers

When there are many layers in a map document, you may want to restrict which ones are selectable. That simplifies the selection process and avoids selecting multiple features from different layers.

1 Clear selected features and click Bookmarks > 48 Contiguous States.

2 In the table of contents, turn off layers COCounties and COStreets.

3 At the top of the table of contents, click the List By Selection button.

4 Clear the check boxes for COStreets, COCounties, and USStates. Now only US Cities will be selectable.

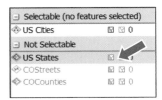

5 Click the Select Features button and click a city. The selected city gets the selection symbol and color that you chose on the previous page and is listed in the table of contents.

1-1
1-2
1-3
1-4
1-5
1-6
1-7
1-8
1-9
A1-1
A1-2

6 Clear the selected features.

Select by graphic

Selecting features by using graphics is a shortcut to select multiple features.

1 Click Bookmarks > Florida.

2 Click the list of the Select Features button and click Select by Circle.

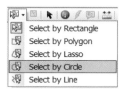

3 Click inside the state of Florida and drag to draw a circle that includes the three cities in Florida.

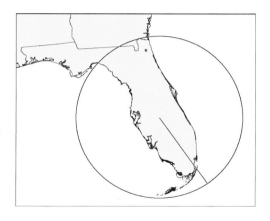

4 Click Bookmarks > 48 Contiguous States. The resulting map shows multiple cities selected and the resulting names in the table of contents.

5 Save your map document.

Tutorial 1-8

Working with attribute tables

You can view and work with data for map features in the layer's attribute table.

Open the attribute table of the US Cities layer and select a record

To explore the attributes of a layer on a map, open its attribute table and select a feature.

1 Save your map document as **Tutorial1-8.mxd** to the Chapter1 folder.

2 In the table of contents, right-click US Cities and click Open Attribute Table. The table opens, containing one record for each US City point feature. Every layer has an attribute table with one record per feature.

3 Scroll down in the table until you find Chicago and click the record selector (gray cell on the left side of the row) for Chicago to select that record. If a feature is selected in the attribute table, it also is selected on the map.

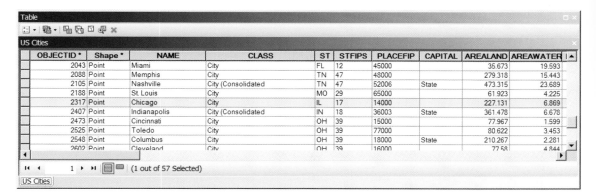

4 Resize the US Cities table to see both the map and table on the screen. You should see the city of Chicago on the map.

5 In the table, click the Clear Selection button ⊡.

Select features on the map and see selected records

1 Select all cities in Florida on the map.

2 In the US Cities table, click the Show Selected Records button. This shows the records for only the features selected in the map: the cities in Florida.

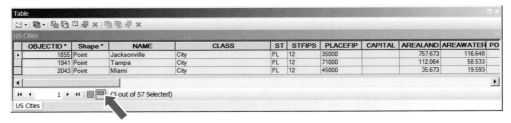

3 Click the Show All Records button 📊 to show all records again. The Florida cities will still be selected.

Switch selections

You can select records in a layer and reverse the selection.

1 In the US Cities table, click the option of the Table Options button ▤ ▾.

2 Click Switch Selection. This reverses the selection and selects all records that were not selected (cities outside of Florida) and deselects those that were selected (Florida cities).

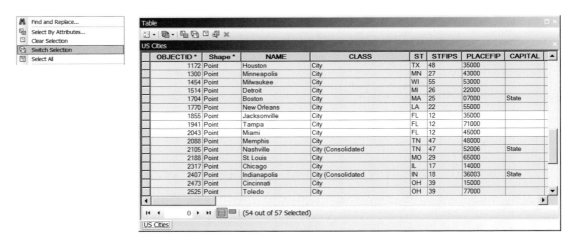

3 Clear the selected features.

Move a field

1 In the US Cities table, click the gray title of the POP2000 field.

2 Drag the POP2000 field to the right of the NAME field.

US Cities						
OBJECTID *	Shape *	NAME	POP2000	CLASS	ST	STFIP ▲
27	Point	Honolulu	371657	Census Designated Place	HI	15
88	Point	Seattle	563374	City	WA	53
130	Point	Portland	529121	City	OR	41
283	Point	Oakland	399484	City	CA	06
287	Point	San Francisco	776733	City	CA	06
325	Point	San Jose	894943	City	CA	06
377	Point	Sacramento	407018	City	CA	06
400	Point	Fresno	427652	City	CA	06
425	Point	San Diego	1223400	City	CA	06
469	Point	Los Angeles	3694820	City	CA	06
517	Point	Long Beach	461522	City	CA	06
560	Point	Anaheim	328014	City	CA	06
564	Point	Santa Ana	337977	City	CA	06
650	Point	Las Vegas	478434	City	NV	32
664	Point	Tucson	486699	City	AZ	04
696	Point	Phoenix	1321045	City	AZ	04

Sort a field

1 In the US Cities table, right-click the NAME field name and click Sort Ascending. This sorts the table from A to Z by the name of each US city with population over 300,000 in year 2000.

US Cities						
OBJECTID *	Shape *	NAME	POP2000	CLASS	ST	STFIP ▲
814	Point	Albuquerque	448607	City	NM	35
560	Point	Anaheim	328014	City	CA	06
879	Point	Arlington	332969	City	TX	48
2755	Point	Atlanta	416474	City	GA	13
1128	Point	Austin	656562	City	TX	48
3032	Point	Baltimore	651154	City	MD	24
1704	Point	Boston	589141	City	MA	25
2833	Point	Charlotte	540828	City	NC	37
2317	Point	Chicago	2896016	City	IL	17
2473	Point	Cincinnati	331285	City	OH	39
2602	Point	Cleveland	478403	City	OH	39
797	Point	Colorado Springs	360890	City	CO	08
2548	Point	Columbus	711470	City	OH	39
900	Point	Dallas	1188580	City	TX	48
790	Point	Denver	554636	City	CO	08
1514	Point	Detroit	951270	City	MI	26

2 Right-click the POP2000 field and click Sort Descending. This sorts the field from the highest populated city to the lowest populated city in year 2000.

YOUR TURN

Move the POP2007 field to the right of POP2000. Sort this field in descending order.

Use Advanced Sorting

1   In the US Cities table, move the ST (State) field to the left of the POP2000 field.

2   Right-click the ST field and click Advanced Sorting.

3   Make selections as follows:

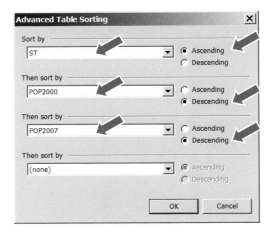

4   Click OK. This sorts the table first by state and then by population of each US city for years 2000, then 2007 for that state.

> **YOUR TURN**
>
> Move and sort by other field names. Try sorting by other multiple fields. For example, you could sort US Cities alphabetically or by state capitals.

Get statistics

You can get descriptive statistics, such as the mean and maximum value of an attribute, in ArcMap by opening a map layer's attribute table using the Statistics function.

1 Close the table and click Bookmarks > 48 Contiguous States.

2 In the table of contents, right-click US States, click Selection, and click Make This The Only Selectable Layer.

3 Hold down the SHIFT key and use the Select Features tool to select the state of Texas and the four states adjacent to it.

4 In the table of contents, right-click US States and click Open Attribute Table.

5 Scroll to the right and right-click the column heading for the POP_2010 field and click Statistics. The resulting window has statistics for the five selected states; for example, the mean 2010 population is 7,681,076.

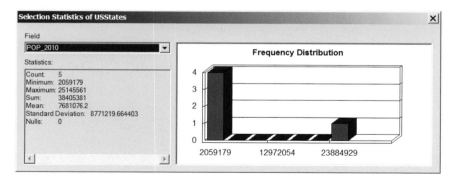

YOUR TURN

Get statistics for the same states in 2000. This should be 6,652,779. Get population statistics for a new selection of states or another field of your choice. Fields in this table are from the 2010 Census.

Turn fields off

Some attribute tables will have many fields and the attribute table can be confusing. You can easily turn fields on or off. If you export a table with a field off it will be excluded from the resulting saved table. In this exercise you will turn on only fields related to population and race.

1 In the table of contents, right-click US States and click Properties.

2 Click the Fields tab and the Turn all fields off button ▣.

3 Turn on the fields ST_NAME, WHITE, BLACK, AMER_ES, ASIAN, OTHER, HISPANIC, and POP2010 and click OK.

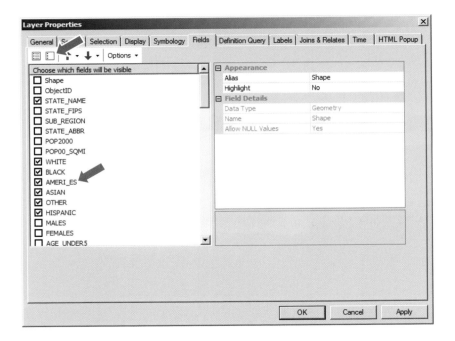

4 In the table of contents, right-click US States and click Open Attribute Table. The table should display population and race data only for 2010.

5 Close the attribute table.

> **YOUR TURN**
>
> Turn off all fields except population 2010 and race for COCounties. Save your map document.

Tutorial 1-9

Labeling features

Labels are text items on the map derived from one or more feature attributes. ArcMap places labels dynamically depending on map scale.

Set label properties and label features

There are many label properties that you can set. Here you get started by setting the data value source.

1   Save your map document as **Tutorial1-9.mxd** to the Chapter1 folder.

2   Close all tables and clear selected features.

3   Click Bookmarks > Florida.

4   In the table of contents, right-click US Cities, click Properties > Labels tab.

5   Click the Label Field arrow and click NAME if it is not already selected.

6   Click the Symbol button > Edit Symbol > and the Mask tab > Halo.

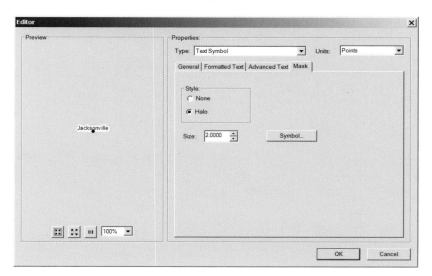

7   Click OK > OK.

8   Select the check box beside Label features in this layer and change the font to bold.

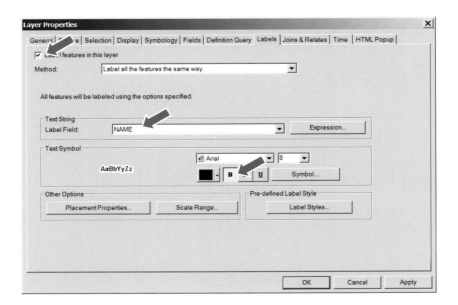

9 Click OK.

10 Zoom out to see additional states.

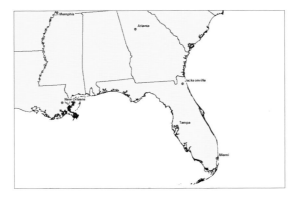

Turn labels off

1 In the table of contents, right-click US Cities and click Label Features.

2 Click Label Features again to turn them back on.

Convert labels to annotation

You can convert labels to graphics in order to edit them individually. You can convert all labels, only labels in a zoomed window, or labels from selected features only.

1 Click Bookmarks > Florida.

2 In the table of contents, right-click US Cities and click Convert Labels to Annotation.

3 Make selections as shown in the image on the right to label features in the state of Florida only.

4 Click Convert.

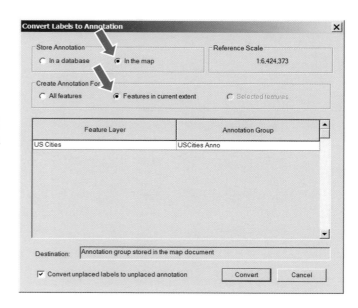

1-1
1-2
1-3
1-4
1-5
1-6
1-7
1-8
1-9
A1-1
A1-2

Edit a label graphic

Once labels become graphics, you can move, scale, and otherwise change them individually.

1 On the Tools toolbar, click the Select Elements button.

2 Click the text label for Miami and move it into the state of Florida.

3 Similarly move the label for Jacksonville.

4 Save your map document and close ArcMap.

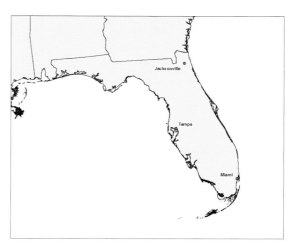

Assignment 1-1

Analyze population by race in the top 10 US states

In this assignment, you will generate statistics for selected race fields for the 10 most populous US states and territories to generate a report with a map graphic and table generated from the GIS attribute table and Microsoft Excel. In chapter 4 you will learn how to calculate data directly in an ArcGIS table. In chapter 5 you will learn how to download census data as used in this assignment.

Get set up

First, rename your assignment folder and create a map document.

- Rename the folder \EsriPress\GIST1\MyAssignments\Chapter1\Assignment1-1YourName\ to your name or student ID. Store all files that you produce for this assignment in this folder.
- Create a new map document called **Assignment1-1YourName.mxd** with relative paths.

You'll learn about map projections in chapter 5. Here, use the following steps to project the table of contents layer to a common projection for the 48 contiguous states:

- Right-click Layers in the table of contents, click Properties > Coordinate System tab.
- Expand Projected Coordinate Systems > Continental > North America and click North America Albers Equal Area Conic > OK.

Build the map

Add the following to your map document:

- \EsriPress\GIST1\Data\UnitedStates.gdb\USStates—polygon features of US states and territories with Census 2010 data. An attribute is STATE_ABBR = two-letter state abbreviation.

Requirements

- Symbolize the layer with a hollow fill and a medium-gray outline (30%).
- Zoom to the 48 contiguous states and create a spatial bookmark.
- Label states with their two-letter abbreviation using a yellow halo mask label. **Hint:** To create a halo label, right-click Layers in the table of contents, click Properties > Labels. Click the Symbol button, Edit Symbol, and the Mask tab. Click Halo, the Symbol button, and yellow as the symbol color.
- Change the selection color to Mars Red and use the USStates attribute table to select the 10 US states and territories having the highest population in 2010.
- Click File > Export Map and save your map as **Assignment1-1YourName.jpg** with 150 dpi resolution.

Get statistics

Using the USStates attribute table, select the top 10 states and territories in year 2010 by population to generate the first data column (Population by race: 10 most populous states and territories) in a Microsoft Excel spreadsheet similar to the following table. Use the same attribute table without selections to create the second data column (Total population by race: All states and territories). Finally, use an Excel expression to create the third data column (Percentage of population by race: 10 most populous states and territories). If you have trouble using Excel you can create a table in Microsoft Word, manually enter the race data, and use a calculator for the percentage data.

Get the statistics from the ArcMap document. In the attribute table, select a statistic in the Statistics output table, press CTRL+C to copy the statistic, click in the appropriate cell of your Excel or Word table, and press CTRL+V to paste it.

In Excel, format cells for the first two data columns to use a 1000 separator (,) and zero (0) decimal places. Use the expression "= first data column/second data column *100" to get the third percentage column where you substitute appropriate cell addresses for the first and second columns.

	Population by race: 10 most populous states and territories	Population by race: All states and territories	Percentage of population by race: 10 most populous states and territories
White			
Black			
American Indian/ Eskimo			
Asian			
Hawaiian-Pacific Islander			
Other			
Total:			

Create a Word document

Create a Microsoft Word document called **Assignment1-1YourName.docx** and include the following:

- Title
- Your name
- Paragraph mentioning each exhibit that follows and any patterns that you observe.
- Exported map inserted—include a caption at the top such as "Exhibit 1. Top 10 most populous states and territories, year 2010."
- Excel or Word table pasted—include a caption at the top such as "Exhibit 2. Statistics on populations by race."

Assignment 1-2

Produce a crime map

Crime prevention depends to a large extent on "informal guardianship," meaning that neighborhood residents keep an eye on suspicious behavior and, if necessary, intervene in some fashion, such as calling the police. Neighborhood crime-watch groups enhance guardianship, so police departments actively promote and support them and keep them informed on crime trends. Suppose that a police officer of a precinct has a notebook computer and projector for use at crime watch meetings. Your job is to get the officer ready for a meeting with the 100 Block Erin Street crime watch group of the Middle Hill neighborhood of Pittsburgh, Pennsylvania.

Get set up

- Rename the folder \EsriPress\GIST1\MyAssignments\Chapter1\Assignment1-2YourName\ to your name or student ID. Store all files that you produce for this assignment in this folder. Create a new map document called **Assignment1-2YourName.mxd** using relative paths.

Build the map

Add the following to your map document:

- \EsriPress\GIST1\Data\Pittsburgh\Midhill.gdb\Bldgs—polygon features for buildings in the Middle Hill neighborhood of Pittsburgh.
- \EsriPress\GIST1\Data\Pittsburgh\Midhill.gdb\CADCalls—point features for 911 computer-aided dispatch police calls in the Middle Hill neighborhood of Pittsburgh. Attributes include NATURE_COD = call type, CALLDATE = date of incident, and ADDRESS = addresses of incident location.
- \EsriPress\GIST1\Data\Pittsburgh\Midhill.gdb\Curbs—line features for curbs in the Middle Hill neighborhood of Pittsburgh.
- \EsriPress\GIST1\Data\Pittsburgh\Midhill.gdb\Streets—line features for street centerlines in the Middle Hill neighborhood of Pittsburgh. Attributes include FNAME = street name, LEFTADD1 = beginning house number on the left side of the street, LEFTADD2 = ending house number on the left side of the street, RGTADD1 = beginning house number on the right side of the street, and RGTADD2 = ending house number on the right side of the street.

Requirements

- Display buildings and curbs as medium-light gray (20%), CADCalls as bright red circles, and streets as *no color* (this is a "trick" to use street names from this layer for curb labels). Label street names using the FNAME field. Create a spatial bookmark of the zoomed area of 100 Erin Street. The 100 block of Erin Street is the segment of Erin Street where addresses range from 100 to 199 and is perpendicular to Webster and Wylie streets.

- Set CADCalls as the only selectable layer and select the crimes in the blocks on either side of Erin Street between Webster, Wiley, Davenport, and Trent. Although it appears that there are only six incident points, there are actually 13 total because multiple incidents are at the same locations. Create a new layer of these crimes called Erin Street Crimes using a blue square. Clear your selected features.
- Click File > Export Map and save your map zoomed to Erin Street as **Assignment1-2YourName.jpg** with 150 dpi resolution.
- For attributes of the Erin Street Crimes layer, turn on fields NATURE_COD, ADDRESS, and CALLDATE only. In the Attributes of Erin Street Crimes table, click the Table Options button and "Export" and save the selected records to a dBASE (.dbf) file called **Assignment1-2YourName.dbf**. Open the dBASE file in Microsoft Excel. When opening the dBASE file in Excel, from the Files of Type menu, select All Files (*.*). This will allow you to choose the dBASE file. Change the column headings to CRIME TYPE, ADDRESS, and DATE. Save your new spreadsheet as **Assignment1-2.xlsx**.

Create a Microsoft PowerPoint document

Create a PowerPoint presentation called **Assignment1-2YourName.pptx** that includes the following:

- Title slide including your name
- Slide with map of Erin Street Crime Block Watch including the exported map zoomed to Erin Street
- Excel table listing the crime type, address, and date.

Part I

Using and making maps

2

Map design

In this chapter you learn all steps necessary to symbolize common types of maps. Most analytical maps convey information with polygons or points using either qualitative or numeric attributes. Sometimes, but less often, lines are the primary interest. For polygons, symbolization mostly uses color fill and boundary lines. For points, it's size, shape, color fill, and boundary lines of point markers. For lines, it's type of line (for example, solid or dashed), width, and color. Symbolization is easy to implement because ArcMap uses attribute values to automate drawing; for example, it can draw all food pantry facilities in a city with a black square point marker of a certain size and all soup kitchen facilities with a black circle of a certain size by using an attribute with type of facility code values.

Learning objectives

- *Symbolize maps using qualitative attributes*
- *Symbolize maps using quantitative attributes*
- *Create custom numeric classes*
- *Create normalized and density maps*
- *Create dot density maps*
- *Create fishnet maps*
- *Create group layers*

Tutorial 2-1

Creating point and polygon maps using qualitative attributes

Placing objects of all kinds into meaningful classes or categories is a major goal of science. Classification in tabular data is accomplished using attributes with codes that have mutually exclusive and exhaustive qualitative values. For example, a code for size could have values "low," "medium," and "high." Any instance of the features with this code is displayed in only one of the classes (the values are mutually exclusive) and there are no more size classes (the values are exhaustive). In this tutorial you learn how to symbolize mapped features—points, lines, and polygons—by class membership as available in code attributes.

Start ArcMap and open a map document

1 On the taskbar, click Start > All Programs > ArcGIS > ArcMap 10.1.

2 Open Tutorial2-1.mxd from the Maps folder. ArcMap opens a map with no layers added. You add the needed layers next.

3 Save the map document to the Chapter2 folder of MyExercises.

Add and display polygons using a single symbol

The first layer of your map is an outline of New York City neighborhoods used as a reference layer.

1 Click the Add Data button, browse through the Data folder to NYC.gdb, click Neighborhoods > Add. ArcMap draws the neighborhoods for New York City as filled polygons.

2 In the table of contents, double-click the polygon symbol and select Hollow symbol, select an outline width of 1.15, and select an outline color of Black.

2-1
2-2
2-3
2-4
2-5
2-6
2-7
2-8
A2-1
A2-2
A2-3

3 In the table of contents, right-click the Neighborhoods layer, click Properties > Label tab.

4 Select Label features in this layer, in Label Field select Name, and select 10 as the Text Symbol size.

Next, you add a white halo to the label to make the labels stand out and be more legible.

5 Click the Symbol button > Edit Symbol button > Mask tab > Halo option button.

6 Type **1.5** for Size and click OK > OK > OK.

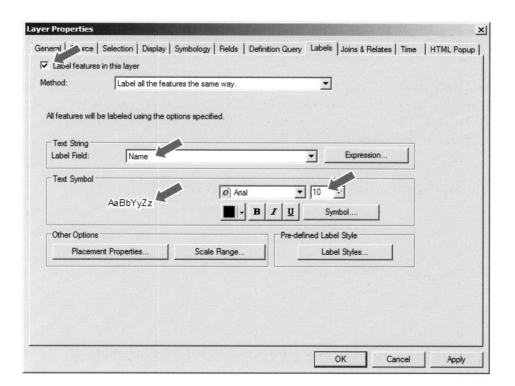

7 Click Bookmarks > Lower Manhattan. ArcMap zooms to this neighborhood and a few neighborhoods in the surrounding boroughs.

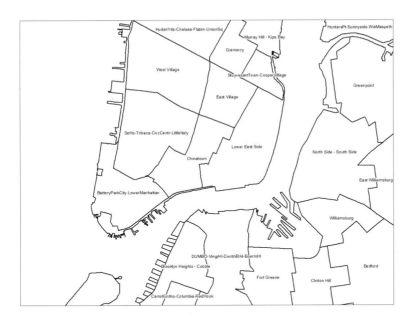

YOUR TURN

Add the Water feature class from NYC.dbf with a blue symbol and no outline. Label the layer using the field LANDNAME with Times New Roman, Bold, and Italic font of size 12. Drag this layer below Neighborhoods. The two layers you added are from different sources so boundaries do not match perfectly.

Add and display polygons using unique symbols

The next layer you add has zoning polygons with various classes of land uses. Zoning restricts land uses to approved kinds; for example, areas zoned residential cannot have businesses located within them.

1 Click the Add Data button, browse through the Data folder to NYC.gdb, click ZoningLandUse > Add.

2 In the table of contents, right-click the ZoningLandUse layer, click Properties > Symbology tab.

3 In the Show panel, click Categories > Unique Values.

4 Under Value Field, click LANDUSE and click Add All Values.

ArcMap assigns colors unique to each land use. Next, you assign colors that are commonly used in zoning maps.

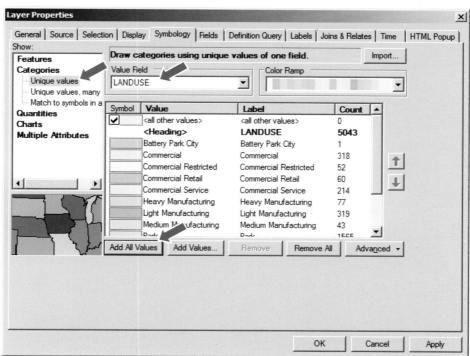

5 Double-click the symbol beside Battery Park City, click Fill Color, and choose color Apple Dust (column 6, row 7 of the color palette) and Gray 30% as outline color. Click OK.

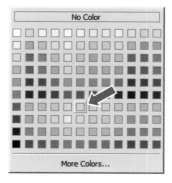

6 Assign a Gray 30% outline to the remaining symbols using the following colors (with column number, row number of the color palette included in parentheses):

- Commercial, Rose Quartz (2,1);
- Commercial Restricted, Medium Coral (2,8);
- Commercial Retail, Tulip Pink (2,9);
- Commercial Service, Rose Dust (2,7);
- Heavy Manufacturing, Blackberry (11,10);
- Light Manufacturing, Lepidolite Lilac (11,1);
- Medium Manufacturing, Lilac Dust (11,7);
- Park, Medium Apple (7,3) ;
- Residential, Yucca Yellow (5,1);
- Residential/Lt Mfg, Light Sienna (4,9), and
- Waterfront, Blue Gray Dust (9,7).

7 Click Apply > OK.

Label zoning features

1 In the table of contents, right-click the ZoningLandUse layer, click Properties > Labels tab.

2 Click Label features in this layer, select ZONE as the Label field, 6 as the Text Symbol size, Gray 60% as the color, and click OK. The result is a subtle label for zoning details.

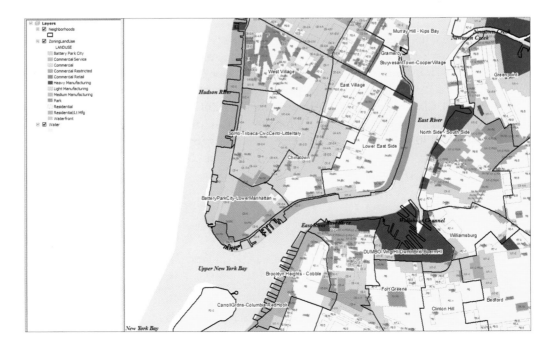

> *YOUR TURN*
>
> ArcMap has predefined symbols for some zoning and land-use features. To view these, double-click the Park symbol > Style References > Civic > OK. Scroll through the Civic polygon symbols and choose Park & Open Space. When finished, deselect Civic under Style References.

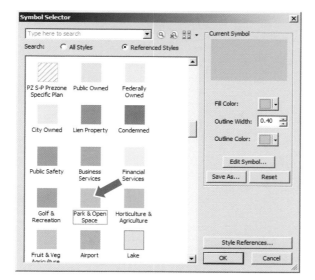

Add points and create a definition query

Often a map layer has more classes of features than you wish to display. In this case you can use a definition query to select the desired subset from the larger collection. Below you add a new layer to your map document, Facilities, which has various government and nonprofit facilities that provide services. Needed for the map are only three classes out of over 100. Classes have both a numeric code (Facility_T) and a corresponding text code (Facttype__1), and the three needed facility classes are 4901 = Soup Kitchen, 4902 = Food Pantry, and 4903 = Joint Soup Kitchen and Food Pantry. While in common language, we use "and" to connect members of a collection such as the three food facility types, in a query you have to use "or."

1 Click the Add Data button, browse through the Data folder to NYC.gdb, click Facilities > Add.

2 In the table of contents, right-click the Facilities layer, click Properties > Definition Query tab.

3 Click the Query Builder button.

4   In the Query Builder window, double-click "FACILITY_T".

5   Click = as the logical operator.

6   Click Get Unique Values. The resulting list has all unique values in the FACILITY_T attribute.

7   In the Unique Values list, double-click 4901.

8   Click OR, double-click "FACILITY_T", click = as the logical operator, and double-click 4902.

9   Repeat step 8 except use 4903 instead of 4902. The completed query, "FACILITY_T" = 4901 OR "FACILITY_T" = 4902 OR "FACILITY_T" = 4903, yields a layer with only food type facilities. If your query has an error, edit it in the lower panel of the Query Builder, or click Clear and repeat steps 4 through 8.

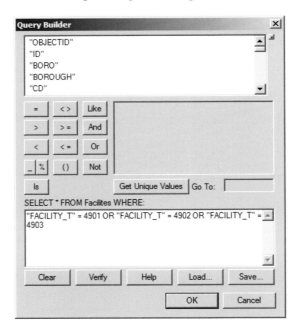

10   Click OK > OK to execute your query. Close the Layer Properties window.

Display points using unique symbols

Next you symbolize food facilities using unique symbols.

1 In the table of contents, right-click the Facilities layer, click Properties > Symbology tab.

2 In the Show panel, click Categories > Unique Values.

3 Click Factype__1 as the Value field > Add All Values.

ArcMap assigns random unique symbols for each food facility type. Next, you assign three specific unique symbols.

4 Double-click the symbol beside Food Pantry and select Esri symbol Square 1 and size 12.

5 Double-click the symbol beside Joint Soup Kitchen and Food Pantry and select Esri symbol Cross 2, size 12. Click OK.

6 Double-click the symbol beside Soup Kitchen and select Esri symbol Circle 1, size 12.

7 Clear the check box beside <all other values> and click OK.

2-1
2-2
2-3
2-4
2-5
2-6
2-7
2-8
A2-1
A2-2
A2-3

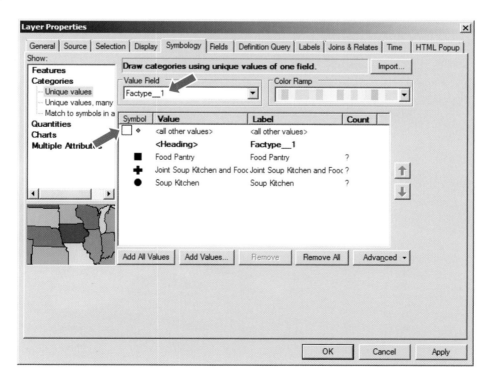

8 Rename the Facilities layer **Food Facilities** and label using the field FACILITY_N, size 6 font, Fir Green text, and a white halo.

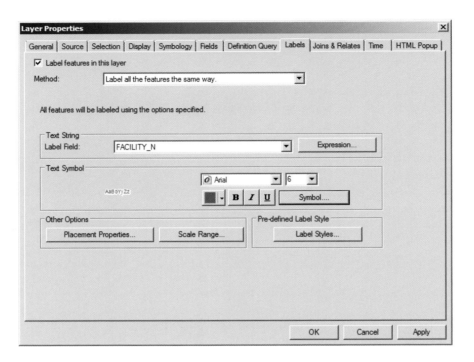

Add and display lines using a single symbol

2-1
2-2
2-3
2-4
2-5
2-6
2-7
2-8
A2-1
A2-2
A2-3

The final layer of your map is street centerlines, which is used as a reference layer. The street layer for all of New York City is very large, so to save space you add streets for Manhattan only.

1 Zoom to the Lower East Side neighborhood.

2 Click the Add Data button, browse through the Data folder to NYC.gdb, click ManhattanStreets > Add.

3 In the table of contents, double-click the line symbol and select a color Gray 40%.

4 In the table of contents, right-click the ManhattanStreets layer, click Properties > Labels tab.

5 Click Label features in this layer, select Street as the Label field, 6 as the Text Symbol size, Gray 70% as the color. Click Placement Properties > Conflict Detection, select Low for Label Weight, and click OK > OK.

6 Save your map document.

Tutorial 2-2

Creating point and polygon maps using quantitative attributes

You cannot show continuous variation in points or polygons; the human eye simply cannot make distinctions unless there are relatively large changes in graphical elements. Generally you have to break a numeric attribute up into classes, similar to what you do to create a bar chart for a numeric attribute. Each bar has a minimum and maximum attribute value for its sides. By tradition the minimum value is included in the bar but the maximum goes in the next bar to the right. All that is needed to symbolize map features is the set of maximum values for classes, which are called "break values."

Open a map document

1 Open Tutorial2-2.mxd from the Maps folder. ArcMap opens a map with US Cities, US States, and USCounties layers added. You use these layers to create choropleth and graduated point maps.

2 Save the map document as **Tutorial2-2.mxd** to the Chapter2 folder.

Create a choropleth map of vacant housing units by county

A choropleth map uses color fill in polygons to represent numeric attribute values. Generally, increasing color value, or darkness of a color fill, in a color ramp represents either increasing or decreasing values. In this exercise, you will use US Census data to create choropleth maps for vacant housing units by county using the quantile classification method. Natural breaks (Jenks) is the default classification method that ArcMap uses but you rarely use this method. Natural breaks uses an algorithm to cluster values of the numeric attribute into groups, with the boundaries of the groups defining classes. You typically choose other methods for classifying your data, including your own custom classification, which you will do in tutorial 2-3.

1 In the table of contents, right-click US Counties, click Properties > Symbology tab.

2 In the Show panel, click Quantities > Graduated colors.

3 Click VACANT as the Fields Value. This is the total number of vacant housing units by county from the 2010 census. The labels reflect the values in the VACANT column.

4 Click the white to black color ramp. This is the sixth color ramp from the bottom.

5 Click the Classify button, Quantile as the Classify Method, 5 as the number of classes, and OK.

6 Click Label and Format Labels.

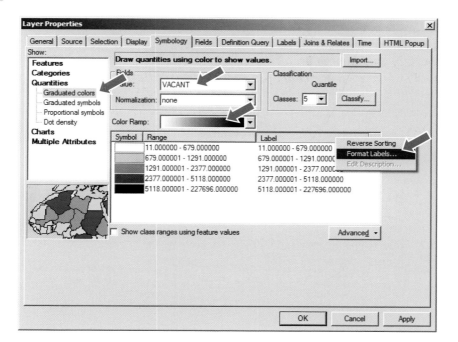

7 Click Numeric under Category, type **0** for the Number of decimal places rounding, and select the check box beside Show thousands separators.

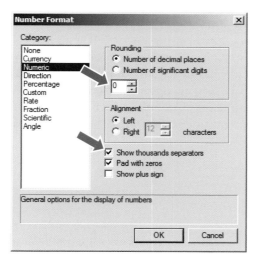

8 Click OK > OK.

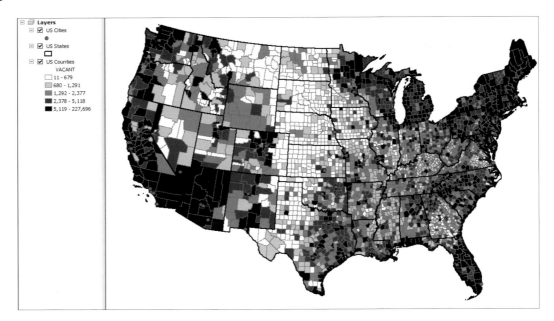

Create a graduated point map of US cities by population

Point maps show locations of data or events using point markers. As a mathematical object, a point has no area, but to see a point on a map you have to use a point marker with area, where the point is at the center of the marker. In the next exercise, you create a point map showing US major cities (2007 population over 250,000) as points with size-graduated point markers. The larger the marker, the higher the population.

1 In the table of contents, right-click US Cities, click Properties > Symbology tab.

2 In the Show panel, click Quantities > Graduated symbols.

3 Click POP2007 as the Fields Value.

4 Click symbol under Template, choose symbol Circle 2, Mars Red as the color, and click OK.

5 Click the Classify button, Quantile as the Classify Method, 5 as the number of classes, and OK.

6 Click the Label title, Format labels, and Numeric under Category.

7 Type **0** for the Number of decimal places rounding and select the check box beside Show thousands separators. Click OK.

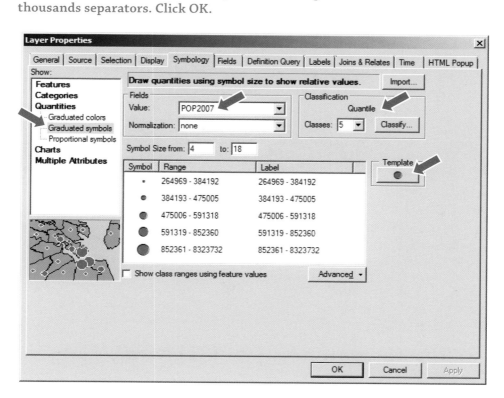

2-1
2-2
2-3
2-4
2-5
2-6
2-7
2-8
A2-1
A2-2
A2-3

8 Click OK. The map shows US Cities with low to high populations compared to US Counties with the highest number of vacant housing units. Do cities with high population appear to have the highest vacant housing units? A custom map will help answer this question.

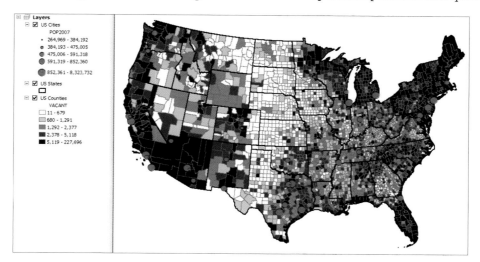

9 Save your map document.

Tutorial 2-3

Creating custom classes for a map

You often need to create your own custom classifications. For example, if the frequency distribution of the attribute is long-tailed to the right (generally has low values but has a set of high values), then a good classification scheme has increasing width intervals in going from low to high values. That allows you to provide detailed information for the bulk of low values, but also have classes for the high values.

Create a custom classification

1 Save your map document as **Tutorial2-3.mxd** to the Chapter2 folder.

2 In the table of contents, right-click the US Counties layer, click Properties > General tab. Rename the layer **Vacant Housing by County**.

3 Click the Symbology tab and the Classify button.

4 Click 6 for the number of classes and select Manual for classification method.

Next, you create new breaks using increasing interval widths that double with each successive class.

5 In the Break Values panel, click the first value, 601, to highlight it. Type **6,000** and press ENTER.

6 Continue clicking the break values below the previous one changed and enter the following: **12,000**, **24,000**, **48,000**, and **96,000**. Let the last (maximum) value remain 227,696.

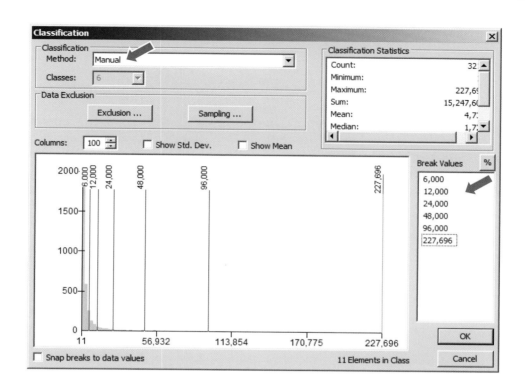

7 Click OK.

8 In the Label field of the Symbology tab, change the first value to **6,000 or less** and the last label to **96,001 or greater**.

Symbol	Range	Label
	11.000000 - 6000.000000	6,000 or less
	6000.000001 - 12000.000000	6,001 - 12,000
	12000.000001 - 24000.000000	12,001 - 24,000
	24000.000001 - 48000.000000	24,001 - 48,000
	48000.000001 - 96000.000000	48,001 - 96,000
	96000.000001 - 227696.000000	96,001 or greater

9 Click OK.

YOUR TURN

Manually change the classes and labels for the 2007 population points to **500,000 or less**, **500,001–1,000,000**, **1,000,001–2,000,000**, **2,000,001–4,000,000**, and **4,000,001 or greater**. Start by clicking value 852360 to change it to 4,000,000 first, and then change the values above it sequentially.

Save a layer file

You can save symbolization of a layer as a layer file (.lyr) for reuse. That can save a lot of time and provide consistency between maps. For example, if you have to produce a new map every day to show recent events, it is efficient to simply apply a layer file to new data, rather than go through all the interactive steps to symbolize from scratch each day.

1 In the table of contents, right-click Vacant Housing by County > Save As Layer File.

2 Browse to the Chapter2 folder of MyExercises, type **Tutorial2-3** in the name field, and click Save.

Import a layer file

Next, you will add the US Counties layer again and apply the custom classes that you saved in the layer file.

1 Click the Add Data button, browse through the Data folder to UnitedStates.gdb, click USCounties > Add.

2 In the table of contents, right-click the US Counties layer, click Properties > General tab.

3 Rename the layer **Owner Occupied by County**.

4 Click the Symbology tab > Import button.

5 Click the browse button, navigate to your Chapter2 folder and click Tutorial2-3.lyr > Add > OK.

6 Click OWNER_OCC as the Value Field > OK > OK.

7 Drag the Owner Occupied by County layer to the bottom of the table of contents and turn the Vacant Housing by County layer off.

2-1
2-2
2-3
2-4
2-5
2-6
2-7
2-8
A2-1
A2-2
A2-3

YOUR TURN

Add the US Counties layer again and use Tutorial2-3.lyr to show Renter Occupied housing units with the same classes as vacant and owner occupied. Rename the layer Renter Occupied by County and drag it to the bottom of the table of contents. When finished, save your map document.

Tutorial 2-4

Creating custom colors for a map

While ArcMap provides color ramps with preselected colors, you can change colors for classes manually. It is best to have more classes with light colors and a few with dark colors because the human eye can differentiate light colors more easily than dark ones. So here you create a custom color ramp that starts with white and ends with a dark blue.

Manually change class colors

1 Save your map document as **Tutorial2-4.mxd** to the Chapter2 folder and turn off the Renter Occupied by County layer.

2 In the table of contents, right-click Owner Occupied by County, click Properties > Symbology tab.

3 Right-click the Color Ramp and click Properties.

4 Click the color box beside Color 1 and click the Arctic White paint chip if it is not already white.

5 Select Color 2, click the color box beside Color 2 > the Dark Navy paint chip > OK > OK. The Population by County map changes to reflect the new color ramp.

Next, you will change each color box manually.

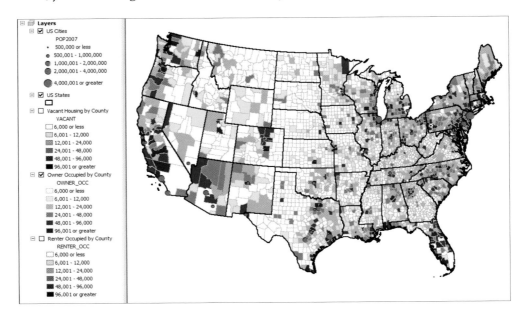

6 In the table of contents, double-click the color box beside 6,001-12,000 for the layer Owner Occupied by County.

7 Click the option beside Fill Color > More Colors.

8 Type **5** for the Saturation (S) value and OK > OK.

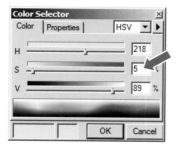

YOUR TURN

Repeat steps 6 and 7 for the next two color boxes of the layer Owner Occupied by County and change the saturation to 20% and 40%. In the Symbology tab of the Layer Properties window, click the Symbol title > Properties for All Symbols, and choose a Gray 30% outline to change the outline color for all color boxes of this layer. Save the layer file as **Tutorial2-4.lyr** and apply the layer with the new colors (and same classes) to the Renter Occupied by County layer. When finished, save your map document. See the graphic on the next page.

2-1
2-2
2-3
2-4
2-5
2-6
2-7
2-8
A2-1
A2-2
A2-3

Tutorial 2-5

Creating normalized and density maps

If you symbolize a choropleth map with population data, you get information useful for doing things such as delivering services for the population. Services require capacities to match the size of the population; for example, a budget, materials, and labor to clean up vacant houses and lots in a city—cut high weeds, remove trash, eliminate vermin, and so forth. A different kind of information is available if you normalize population data. There are two options. One is on the relative sizes of population segments; for example, if you divide (normalize) the population of vacant houses by the total number of houses in a polygon, you get information on how well maintained or poor an area is. Areas with a low percentage of vacant houses are in good shape, while those with a high percentage are in bad shape. The second option is to divide population by the area of polygons, which yields density such as number of vacant houses per square mile. Density is a measure of congestion or concentration. A neighborhood with a high density of vacant houses is much worse off than another neighborhood with a low density. Many vacant houses near each other have a multiplier effect in dragging down a neighborhood.

Create a choropleth map with normalized population

1 Save your map document as **Tutorial2-5.mxd** to the Chapter2 folder, turn on Vacant Housing by County, and turn off Owner Occupied by County and Renter Occupied by County.

2 In the table of contents, right-click the Vacant Housing by County layer, click Properties > General tab.

3 Rename the layer **% Vacant Housing by County**.

4 Click the Symbology tab and select HSE_UNITS as the Normalization field. This will display the fraction of vacant housing units with VACANT as the numerator and HSE_UNITS as the denominator.

5 Click the Classify button, select 5 as the number of classes, Equal Interval for the Classification Method, and click OK.

6 Click the Label title > Format labels > and Percentage under Category.

7   Select the radio button beside "The number represents a fraction."

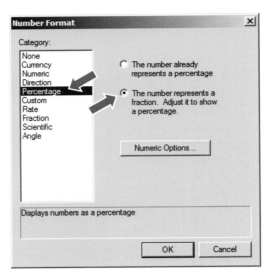

8   Click the Numeric Options button, type **1** as the number of decimal places, and click OK > OK. The following figure shows the final edits in the Symbology tab to show the percentage of vacant housing units.

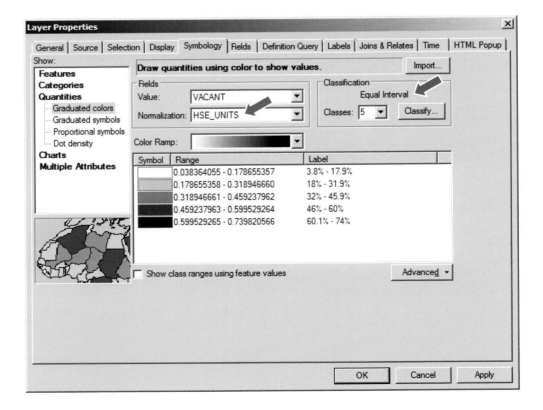

9 Click OK. You now see the US counties with the highest percentage of vacant housing units, a very different map.

Create a density map

The area field used in this exercise, ALAND10, was created by the US Census Bureau and has square meters for units. So the densities you display are vacant houses per square meter.

1 In the table of contents, right-click the % Vacant Housing by County layer > Copy.

2 On the Menu bar, click Edit > Paste.

3 In the table of contents, right-click % Vacant Housing by County, click Properties > General tab. Rename the layer **Vacant Housing Density**.

4 Click the Symbology tab, click ALAND10 as the Normalization field, and click OK.

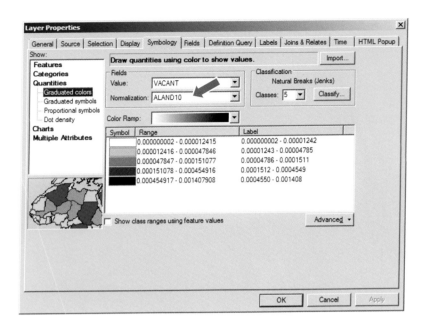

5 In the table of contents, right-click US Cities > Label Features.

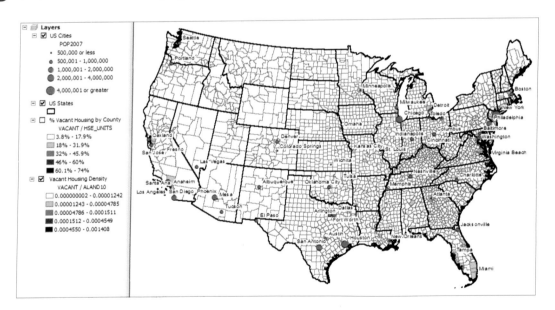

6 Click Bookmarks > New York. Notice the high density of vacant housing units.

2-1
2-2
2-3
2-4
2-5
2-6
2-7
2-8
A2-1
A2-2
A2-3

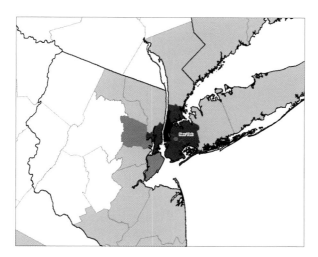

7 Click Bookmarks > Los Angeles. Notice that the density is not as high because the land area of Los Angeles County is much greater than the land area of New York County.

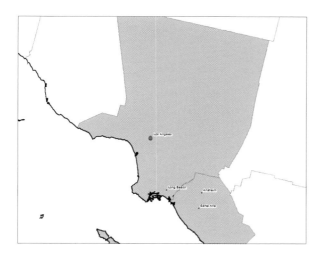

YOUR TURN

Explore the cities under Bookmarks or other cities of interest to you. Click Bookmarks > 48 Contiguous States and save your map document.

Tutorial 2-6

Creating dot density maps

A dot density map is an alternative to the density map of the previous tutorial. Instead of color fill symbolizing polygons, each polygon is filled with randomly placed points based on the value of an attribute, generally a population. For example, each point might represent 100 persons so that a polygon with a population of 1,000 would have 10 randomly placed points in it. The result is a kind of area density map, where the percentage of area on a polygon covered with dots gives you an idea of how densely populated the polygon is.

Open and save the map document

1 Open Tutorial2-6.mxd from the Maps folder. ArcMap opens with a map of counties and states in the United States.

2 Save the map document as **Tutorial2-6.mxd** to the Chapter2 folder.

3 Click Bookmarks > Pennsylvania.

Create a dot density map

1 In the table of contents, right-click the US Population 2010 layer, click Properties > Symbology tab.

2 In the Show panel, click Quantities > Dot density.

3 Click POP_2010 as the Field Selection field and move the field to the right pane.

4 Double-click the dot symbol for POP_2010, click Color > Mars Red > OK.

5 Click the line button in the Background section, select Gray 20% as the color, and click OK.

6 Type **5,000** as the Dot Value.

7 Click OK.

8 Save your map document.

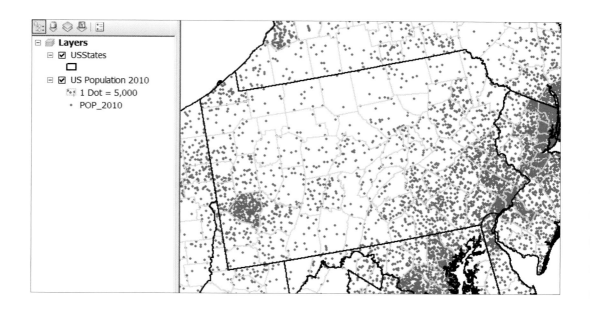

YOUR TURN

Try changing the dot value to 10000 and 100000. Don't save the changes.

Tutorial 2-7

Creating fishnet maps

A fishnet map is so named because its uniform, square polygons (cells) make it look like a fishnet. Here you create a new polygon layer of uniform square cells and then aggregate point data (locations of public schools) to cell summaries (schools per cell). The advantage of a fishnet map is that when symbolized as a choropleth map, all polygon sizes and shapes are identical, so that your perception of variation is driven only by color value. A choropleth map of irregular polygons, such as for counties, has the confusing nature of having two graphic elements that vary: color value and size/shape of polygon. Only the color value of polygons is valid, while large-size, high-color value polygons catch more of your attention, but polygon size has no direct meaning for the symbolized attribute.

Open and save the map document

1 Open Tutorial2-7.mxd from the Maps folder. ArcMap opens with a map of counties and schools for Pennsylvania.

2 Save the map document as **Tutorial2-7.mxd** to the Chapter2 folder.

Get set up for the fishnet tool

You have to calculate the number of rows and columns for a given cell size to cover the extent of your map. The data folder has an Excel workbook that makes this easy. Cells need to be square and the map has to have projected coordinates (not latitude and longitude). So cell size is simply the length of one side of a cell.

1 Using Computer, browse to the Data > DataFiles folder and double-click FishnetCalculations.xlsx. This worksheet has five inputs shown in a brown font. Cell size is the first, which is the length of a side of a cell. Next are four bounding coordinate values—for Pennsylvania, in this case. Top is the maximum y-coordinate, Bottom is the minimum y-coordinate, Right is the maximum x-coordinate, and Left is the minimum x-coordinate. The coordinates in the worksheet are correct for Pennsylvania. Cell size 30,000 is too large, however, so you will change it to 10,000. First, roughly check the coordinates.

	A	B	C	D	E	F
2						
3	Cell size	30,000	Can be changed to whatever you wish			
4						
5	**Number of rows**					
6	Top	4,726,414				
7	Bottom	4,396,769		Divide Height by cell size		
8	Height	329,645		number cells =	10.98818	
9				rounded up =	11	
10	**Number of columns**					
11	Right	1,041,059		Divide Width by cell size		
12	Left	539,442		number cells =	16.72056	
13	Width	501,617		rounded up =	17	
14						

2-1
2-2
2-3
2-4
2-5
2-6
2-7
2-8
A2-1
A2-2
A2-3

2 Using your cursor and coordinates readout at the lower right of the map window, verify the top, bottom, right, and left coordinates for PACounties. While you may not get the identical coordinates in the workbook, you should be close. For a cell size of 30,000 meters (the units of this map), you would need 11 rows and 17 columns for the fishnet.

3 Type **10000** for the cell size. For this cell size you need 33 rows and 51 columns.

4 Save and close the Excel workbook.

Create the fishnet

1 Click Windows > Search, type **fishnet** in the search textbox, press ENTER, and click Create Fishnet (Data Management).

2 Type or make selections as shown in the next figure. The top, bottom, right, and left coordinates displayed in the Create Fishnet tool is an alternative source for these values instead of direct measurement on the map.

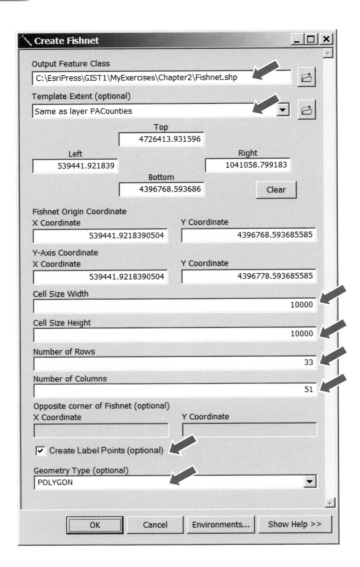

3 Click OK and if necessary close the Search window.

4 Turn off the Fishnet_label layer in the table of contents and symbolize Fishnet with a hollow fill and tan outline color of your choice. The Fishnet_label layer has centroid points for the cells.

5 Move PACounties above the Fishnet in the table of contents.

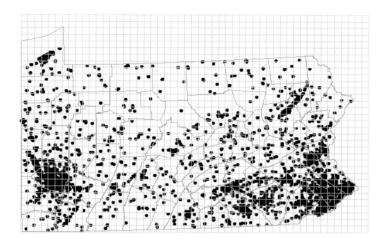

Delete unneeded cells

A number of cells lie outside Pennsylvania's boundary. Next, you eliminate all cells that are completely outside of Pennsylvania.

1 Click Selection > Select By Location and type or make selections as follows:

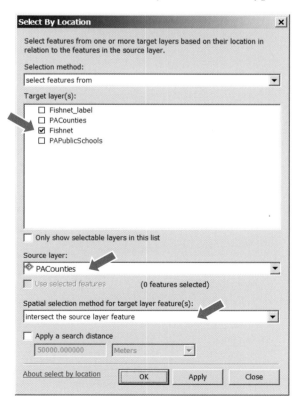

2 Click OK.

3 In the table of contents, right-click Fishnet, click Data > Export Data and type or make selections as follows:

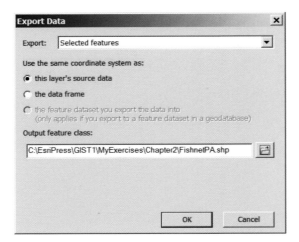

4 Click OK, and Yes to add to the map.

5 Right-click Fishnet_label and click Remove. Do the same for Fishnet.

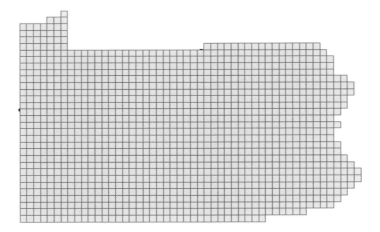

Spatially join schools to the fishnet

This step counts the number of schools per grid cell. With this step completed, you are ready to symbolize a choropleth map for number of schools per grid cell.

1 In the table of contents, right-click FishnetPA, click Joins and Relates > Join, and then type or make selections as follows:

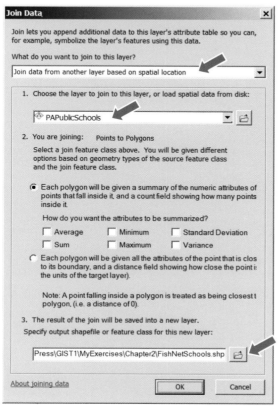

2 Click OK.

3 Remove FishnetPA and PAPublicSchools from the table of contents.

4 In the table of contents, right-click FishnetSchools and click Open Attribute Table.

5 Right-click the Count_ column heading and click Sort Descending. Count_ has the number of schools per grid cell, which has a maximum of 131 for grid cell 216.

6 Close the attribute table and move PACounties above FishnetSchools in the table of contents.

7 In the table of contents, right-click FishnetSchools, click Properties > Symbology.

8 Click Quantities, select Count_ as the value field.

9 Click Classify and select Quantile as the classification method and click OK > OK.

2-1
2-2
2-3
2-4
2-5
2-6
2-7
2-8
A2-1
A2-2
A2-3

10 Resymbolize PACounties to have a darker outline so that it shows up better. The uniform grid makes for a very clear choropleth map.

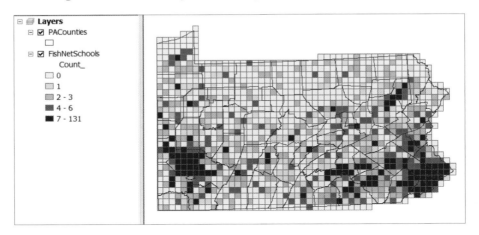

11 Save your map document.

Tutorial 2-8

Creating group layers and layer packages

If you have a map with many layers, some of which are in groups that would be convenient for the user to turn on and off as groups, then it is helpful to create group layers for that purpose. You can also save a group layer (or any single layer) as a layer package (a file with a .lpk extension). A layer package is one file with all data sources and symbolization included, which you can share with others, including online at the Esri website, ArcGIS Online. In this tutorial you create layer groups for populations and facilities in New York City by administrative and political features including fire companies, police precincts, health areas, and school and council districts. You create layer packages from layer groups.

Start ArcMap and begin a map document

1 Click File > New, click Blank Map > OK.

2 Save your map document as **Tutorial2-8.mxd** to the Chapter2 folder.

Create a layer group and description

1 In the table of contents, right-click Layers and click New Group Layer.

2 Right-click the resulting New Group Layer, click Properties.

3 Click the General tab and type **NYC Fire** as the Layer name and **NYC Fire Houses and Population** as the Description. A description is needed to later save the layer group as a layer package.

4 Click the Group tab > Add.

5 Browse through the Data folder to NYC.gdb, hold down the CTRL key, select Boroughs, Facilities, FireCompanies, and Water, and click Add (but do not click OK). These features are to be added to your NYC Fire layer group.

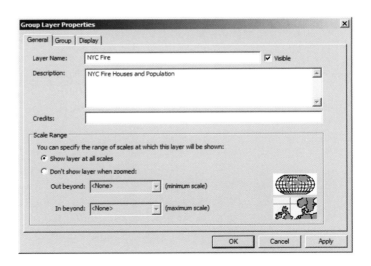

6 Click Boroughs and the up arrow button until it is at the top of the group layer list.

7 Click Water and the down arrow so that it is at the bottom of the group layer list and click OK.

8 Click OK.

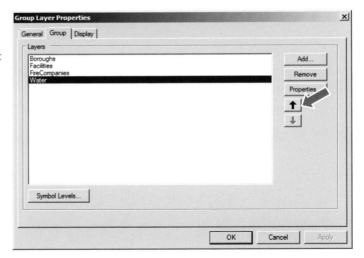

Symbolize layers

1 In the table of contents, right-click Boroughs, click Properties > Symbology tab and choose Single symbol, hollow fill, black outline width 1.5. Leave the properties window open.

2 Click the Labels tab> Label features in this layer checkbox > the Symbol button > the Edit Symbol button > the Mask tab > the Halo option button > OK > OK > OK.

3 In the table of contents, right-click Facilities, click Properties > Definition Query tab and type or select the definition query "**FACILITY_T**" = **2101**. Click OK > OK. This displays just the New York City fire houses. Now you can see the borough boundaries and labels

4 Rename the Facilities layer as **Fire Houses** and display as a Square2, Mars Red, size symbol.

5 In the table of contents, right-click FireCompanies, click Properties > Symbology tab.

6 Click Quantities > Graduated Colors, select Pop2010 as the Value field, select the black to white color ramp, click the symbol heading above the color ramp chips > Flip Symbols (so that you get a white to black color ramp).

7 Click the Classify button, select Quantile as the classification method and click OK, leaving the Layer Properties window open.

8 Click the Label heading over the right half of the symbols, click Format Labels > Show Thousands Separators > OK > OK.

9 In the table of contents, right-click Water, click Properties > Symbology tab and choose Single symbol, Blue, with no outline, and click OK.

You now have a layer group for NYC Fire that includes four layers.

2-1
2-2
2-3
2-4
2-5
2-6
2-7
2-8
A2-1
A2-2
A2-3

YOUR TURN

Create a layer group called **NYC Police** with description **NYC Police Precincts and Population**. Add NYC.gdb layers Boroughs, Facilities, PolicePrecincts, and Water in the same order as the NYC Fire layers. Use the same symbology and labels as the NYC Fire layers. Use Pop2010 as the population field and "FACILITY_T" = 2001 (Police Stations). Symbolize Police Station points as Circle2, blue of your choice, size 6. Rename this layer **Police Stations**. Try turning entire groups on and off by selecting the small checkbox to the left of each group layer name. Save your map document.

Save a layer package

1   In the table of contents, right-click NYC Fire > click Create Layer Package.

2   In the Layer Package window, click Save package to file. ArcGIS should automatically fill in the package path and file name such as C:\EsriPress\GIST1\MyExercises\Chapter2\NYC Fire.lpk

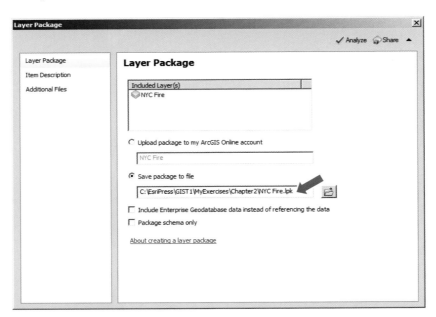

3   Click the Item Description button and type **NYC Fire Houses and Population** as the Summary (required) and **NYC_Fire** as the Tags (required).

4   Click the Share button and wait while the layer package file is created.

5 Click OK when the Succeeded window appears. You now have a layer package that can be shared with another ArcGIS user or uploaded to ArcGIS Online.

YOUR TURN

Create a layer package for Police Precincts. When finished, save your map document and exit ArcMap.

Assignment 2-1

Create a map showing schools in New York City by type

Suppose that a philanthropic foundation wants to study the role of charter schools in K-through-12 education in New York City, including accessibility of students to charter schools. An initial GIS project is simply to map all K-through-12 schools—including public, private, and charter schools.

Get set up

- Rename the folder \EsriPress\GIST1\MyAssignments\Chapter2\Assignment2-1YourName\ to your name or student ID. Store all files that you produce for this assignment in this folder.
- Create a new map document called **Assignment2-1YourName.mxd** with relative paths.

Build the map

Create three group layers called **NYC Public Schools**, **NYC Public Charter Schools**, and **NYC Private Schools**. Include descriptions for each layer group.

Add the following to each layer group in the following order:

- \EsriPress\GIST1\Data\NYC.gdb\Boroughs—polygon features of New York City boroughs.
- \EsriPress\GIST1\Data\NYC.gdb\SchoolDistricts—polygon features of New York City school districts.
- \EsriPress\GIST1\Data\NYC.gdb\Facilities—point features of New York City facilities. Attributes include FACILITY_T = facility type and FacType___3 = Facility type descriptive name.
- \EsriPress\GIST1\Data\NYC.gdb\MetroRoads—line features of New York City major roads.
- \EsriPress\GIST1\Data\NYC.gdb\Water—polygon features of bodies of water in New York City.

Requirements

- Symbolize Boroughs with a hollow fill, black, 2.0 outline width. Label the boroughs using a light yellow halo mask, bold, size 12. Create spatial bookmarks for each borough.
- Symbolize School Districts with a hollow fill, ultra blue, 1.75 outline width. Label the school districts using a white halo mask, bold, size 10 and field SchoolDist.
- Symbolize MetroRoads and Water as ground features and with labels.
- Create the following definition queries for each school type:
- Public Schools "FACILITY_T" >= 1001 AND "FACILITY_T" <= 1006
- Public Charter Schools "FACILITY_T" >= 1011 AND "FACILITY_T" <= 1017
- Private Schools "FACILITY_T" >= 1101 AND "FACILITY_T" <= 1106

- Symbolize each school type using attribute FacType___3 with unique symbols. Use the same colors for elementary, intermediate/JHS, junior/senior, high school, and K-12 schools. Use circles for public schools, squares for public charter schools, and triangles for private schools. Use symbol asterisk 2 for the one special public charter school.
- Save the layer groups as layer packages called **NYCPublicSchools**, **NYCCharterSchools**, and **NYCPrivateSchools**.
- Save the map document with only the NYCPublicSchools layer group turned on and zoomed to Staten Island.

Assignment 2-2

Create maps for military sites and congressional districts

BRAC (Base Realignment and Closure) is a process of the US federal government used by the Department of Defense (DoD) and Congress to close and realign military installations in order to reduce expenditures and achieve increased efficiency. At the same time, members of Congress attempt to keep military installations open in their districts or states as sources of employment and economic activity. More than 350 installations have been closed in five BRAC rounds since 1989 and the most recent round was completed in November 2005. A 2005 commission recommended that Congress authorize another BRAC round in 2015, and then every eight years thereafter. GIS is an effective tool in regard to BRAC to visualize and analyze military installations and congressional districts in the United States.

Get set up

- Rename the folder \EsriPress\GIST1\MyAssignments\Chapter2\Assignment2-2YourName\ to your name or student ID. Store all files that you produce for this assignment in this folder.
- Create a new map document called **Assignment2-2YourName.mxd** with relative paths.

Use the following steps to project the map to a common projection for the 48 contiguous states:

- Right-click Layers in the table of contents, click Properties > Coordinate System tab.
- Expand Projected Coordinate Systems > Continental > North America and click North America Albers Equal Area Conic > OK.

Build the map

Create three group layers: (1) **111th Congress DoD Installation Count**, (2) **DoD Installation Type**, and (3) **Brac Status**. Include descriptions for each layer group.

Add the following to each group layer according to the instructions and rename each layer appropriately. Include the US States layer in every layer group, symbolized to your liking and labeled using the state abbreviations:

- \EsriPress\GIST1\Data\UnitedStates.gdb\USStates—polygon features of US states.
- \EsriPress\GIST1\Data\UnitedStates.gdb\USCities—point features of US cities. An attribute is POP2007 = population of each city in the year 2007.
- \EsriPress\GIST1\Data\UnitedStates.gdb\Congress111—polygon features of generalized boundaries for the 111th congressional districts. Attributes include DoD_COUNT = the number of DoD sites, installations, ranges, and training areas in each congressional district and PARTY (D = democrat, R = republican).
- \EsriPress\GIST1\Data\UnitedStates.gdb\MilitaryBnd—polygon features of 2010 of Department of Defense (DoD) sites, installations, ranges, and training areas in the United

States and Territories. An attributes is COMPONENT = type of military installation (Army, Navy, Air Force, Marine Corps, Washington Headquarters Service).

- \EsriPress\GIST1\Data\UnitedStates.gdb\MilitaryPt—point features of 2010 of Department of Defense (DoD) sites, installations, ranges, and training areas in the United States and Territories. An attribute is BRAC_SITE = site selected for closure under a BRAC action (YES or NO).

Requirements

- Zoom to the 48 contiguous states and create a bookmark zoomed to these states.

- In group layer 1 show the DoD count by congressional district as a choropleth map.

- In layer group 2, show and label US Cities whose populations are greater than or equal to 250,000. Show DoD installations (MilitaryBnd polygons) as unique symbols using the COMPONENT field. Group the component colors based on the types. Use shades of purple for Air Force, green for Army, orange for Marine Corps, blue for navy, and gray for Washington Headquarters Service. When choosing colors, focus on the active installations by making their colors darker than guard or reserve installations.

- In layer group 3, show DoD installations (MilitaryPt points) as unique symbols using the BRAC_SITE field and congressional districts as unique symbols using the PARTY field.

2-1
2-2
2-3
2-4
2-5
2-6
2-7
2-8
A2-1
A2-2
A2-3

Assignment 2-3

Create maps for US veteran unemployment status

The US Census collects data about US veterans and their dependents. In this assignment you create dot density and choropleth maps showing the unemployment status of veterans in the labor workforce by US county. The data for these maps are 2009 three-year estimates from the American Community Survey. Some counties are not populated enough and have no data. You will display these counties as a light gray polygon.

Get set up

- Rename the folder \EsriPress\GIST1\MyAssignments\Chapter2\Assignment2-3YourName\ to your name or student ID. Store all files that you produce for this assignment in this folder.
- Create a new map document called **Assignment2-3YourName.mxd** with relative paths.

Use the following steps to project the table of contents layer to a common projection for the 48 contiguous states:

- Right-click Layers in the table of contents, click Properties > Coordinate System tab.
- Expand Projected Coordinate Systems > Continental > North America and click North America Albers Equal Area Conic > OK.

Build the map

Add the following to your map document:

- \EsriPress\GIST1\Data\UnitedStates.gdb\USStates—polygon features of US states.
- \EsriPress\GIST1\Data\UnitedStates.gdb\USCounties—polygon features of US counties. Attributes include VET_LBRFORCE = number of US veterans in the labor force, VET_ UNEMP = number of unemployed US veterans in the labor force.

Requirements

- Zoom to the 48 contiguous states and make a spatial bookmark called 48 Contiguous States.
- Symbolize USStates with a hollow fill, black, 1.5 point outline. Label every state with its abbreviation using a white halo mask, bold, size 8.

Add the USCounties layer three times, renamed, for the following maps:

- The first is a dot density map using field VET_LBRFORCE and a dark blue dot representing 5,000 veterans in the workforce. Use a light gray (20%) outline for the counties.

- The second is a choropleth map normalized showing the percentage of unemployed veterans in the labor workforce using fields VETUNEMP/VET_LBRFORCE. Use a light to dark red color ramp with 5 classes and classes 0–4%, 4.01–8%, 8.01–16%, 16.01–32%, and 32.01% and greater. Edit the label symbology format to show the numbers as percentages. Use the format "The number represents a fraction" with 2 decimal places.

- The third is a single symbol map at the bottom of the table of contents. Use a light gray (10%) fill color and a medium-light gray outline (30%). Because there are some counties with no veteran employment data, this is a "trick" to show US counties with no data.

2-1
2-2
2-3
2-4
2-5
2-6
2-7
2-8
A2-1
A2-2
A2-3

3

GIS outputs

ArcGIS can produce many forms of output, including interactive maps, printed maps for distribution, image files for use in presentations or on websites, map animations, and maps that you share using the Esri ArcGIS Online cloud services. In this chapter you learn how to build each of these kinds of outputs.

Learning objectives

- *Build an interactive GIS*
- *Build map layouts*
- *Build map animations*
- *Use ArcGIS Online*

Tutorial 3-1

Building an interactive GIS

You already know from chapter 2 how to interact with a map document by zooming in and out, panning, using the Identify tool, and using other tools such as the Magnifier tool. A map document intended for interactive use, however, can have additional enhancements and that's what this tutorial is about. For example, sometimes what you would like to see is highly detailed and cannot be viewed when at full extent, but must be viewed when zoomed in. ArcGIS provides visible scale ranges for map layers for this case, so depending on how far zoomed in or out you are, you see different layers. In addition you need to learn more about labeling features, about a tool called map tips, and about turning map features, such as points, into hyperlinks to websites or files. Finally, to make data clear to the map user, you need to clean up attribute tables, hiding attributes of no use to the map viewer and assigning self-documenting alias names to useful attributes. Note that you need an Internet connection to display a satellite image that is included as a web service in the tutorial map document that you are about to open.

Open a map document

1   **Start ArcMap and open Tutorial3-1.mxd from the Maps folder.** The map document opens with all layers visible at full extent. You can't make out Manhattan streets, which appear to be a brown-filled polygon instead of lines (only Manhattan streets are included instead of all New York City streets to save computer disk space). Also, the point features (food pantries and soup kitchens) are piled on top of each other so that you cannot select individual points for more information. Finally, the satellite image, while interesting to see, does not help with information on where to zoom next. Obscured are polygons for New York City's five boroughs which, if visible and labeled, would be helpful for choosing areas to zoom in to.

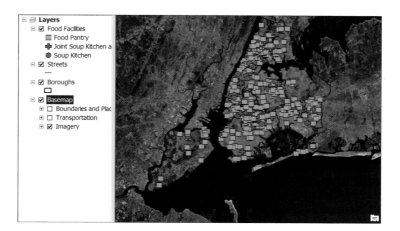

2 Save the map document to the Chapter3 folder of MyExercises. Notice the scale of your map display on the Standard toolbar. The scale in the following image reads 1:500,000 but yours may differ depending on your computer screen size and ArcMap window size. The scale value means that one inch (or foot or any other unit for linear measurement) on the computer screen corresponds to 500,000 inches (or other unit) on the ground. The ratio 1/500,000 is small, so such a map scale is called "small scale." You'll eventually zoom in to the West Side of Manhattan at a scale of approximately 1:30,000, which is a larger fraction and is called "large scale." Obviously, the larger the scale, the more detail you can see and use.

1:500,000	▾

Set visible scale ranges

When you set a minimum scale range for a map layer (the minimum scale at which it displays), the layer does not display if you zoom out any farther than the minimum. When you set a maximum scale range (the maximum scale at which it displays), the layer does not display if you zoom in any farther than the maximum.

1 At the top of the table of contents, click the List By Drawing Order button (the left-most button). This mode of the table of contents provides information on what layers are visible at the current map scale, after you assign visible scale ranges.

Next, you use bookmarks so that it is easy for you to follow instructions and assign visible scale ranges in this tutorial. The bookmarks are not an essential feature for the ultimate map user.

2 Click Bookmarks > Manhattan Borough. In the scale textbox on the Standard toolbar, edit the denominator by adding **1** to it (so, for example, if it is 200,000 make it 200,001) and press the TAB key. The last step, adding 1 to the scale denominator, is a "trick" so that when you use the Manhattan Borough bookmark in the future, the layers that you are about to modify display. Let's turn on the Satellite Image layer at this scale, making its current scale the minimum scale. Then at this scale or zoomed out farther, this layer will not display (but it will appear for the Manhattan bookmark scale, which is one foot farther zoomed in).

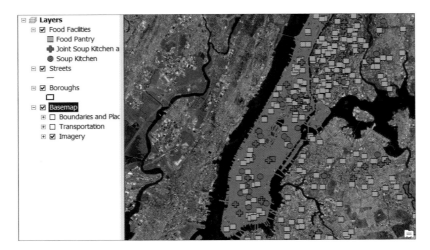

3 Right-click Basemap and click Visible Scale Range > Set Minimum Scale. Do the same for Food Facilities.

4 Click Bookmarks > NYC. Only Boroughs and Streets are on. Notice that the checkmarks for Food Facilities and Basemap are shaded, signifying that these layers are available but not visible at the current scale.

5 Click Bookmarks > Manhattan Borough. All layers are on.

6 Click Bookmarks > Central Park West Side and add 1 to the denominator of the scale, similar to step 2.

7 Set the Minimum Scale Visible Range for Streets. Now try out all of the bookmarks, starting with NYC. NYC now displays just Boroughs. Manhattan Borough adds the Basemap image and Food Facilities. Central Park West Side adds Streets so that all layers are on.

Label food facilities

While food facilities turn on at the scale of the Manhattan Borough bookmark, don't display labels for them until zoomed in farther to the Central Park West Side bookmark.

1 Click Bookmarks > Central Park West Side and add **1** to the denominator of the scale.

2 Right-click Food Facilities, click Properties > Labels. Select the Label features in this layer check box and verify that FACILITY_N is the label field.

First you will give the label a halo to make it highly visible. Then you will set its visible scale range.

3 Click the Symbol button > Edit Symbol > Mask > Halo.

4 Type **1.5** for size and click the Symbol button.

5 Select a pale green for fill color, No Color for outline color, and click OK > OK.

6 In the Symbol Selector window, set size to 6 and click OK.

7 In the Layer Properties window, click the Scale Range button > Don't show labels when zoomed, select <Use current scale> for Out beyond, and click OK > OK.

8 Click Bookmarks > Central Park West Side. You should see labels for Food Facilities. They are very prominent because food facilities is the map layer of primary interest in this map document.

YOUR TURN

Label Streets using Street as the label field with a size 6 font and 1.25 size halo that has a light tan fill color and no outline. For Scale Range (Don't show labels when zoomed Out beyond), use the current scale. In the Labels tab of Properties, click Placement Properties > Placement > Parallel for orientation and On the line for position (turn off Above as a position).

Also, use BoroName as the label field for Boroughs with a size 8 font, 1.5 size halo, a light blue fill for the halo, and no outline.

Finally, for the Central Park West Side bookmark, set a maximum scale for the Boroughs labels. First add 1 to the scale denominator before setting the maximum scale. Then the Manhattan label should not display when at the Central Park West Side bookmark.

Try out all of the bookmarks, starting with NYC. Notice that your street labels have priority over your more important Food Facility labels: not all food facilities get labels. Next, you give Food Facilities labels the preference for display.

Set priorities for labels

1 Open the Labels tab of the Layer Properties for Food Facilities and click Placement Properties > Conflict Detection.

2 Select High for both Label Weight and Feature Weight and click OK > OK.

3 Open the Labels tab of the Layer Properties for Streets and click Placement Properties > Conflict Detection.

4 Select Medium for Label Weight and Low for Feature Weight. Now labels for Food Facilities all display and take priority.

One more enhancement is to reduce the number of street labels created to cut down on clutter, using the next step.

5 Type the value **1** in the Buffer field.

6 Click OK > OK. Now all of the food facilities' labels display plus enough streets to get your bearings.

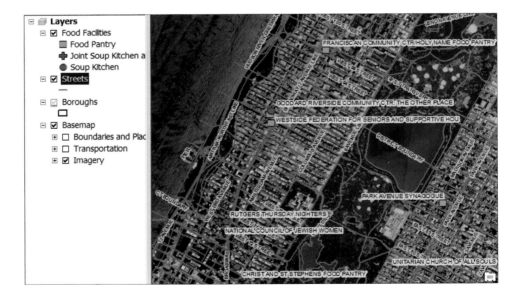

Set field properties

When you use the Identify tool to get information, you should just see relevant attributes and they should have plain English names. Next, you set field properties of Food Facilities and Streets for this purpose.

1 Open the Fields tab of the Layer Properties for Food Facilities and click the Turn Off All Fields button ⊡ (to make them not visible). Select check boxes for FACILITY_N, FACILITY_A, ZIP, and Factype_1 to turn those fields back on (make them visible).

2 Click FACILITY_N to select it, and type **Name** as the Alias. Similarly type **Address** as the alias for FACILITY_A, **ZIP Code** for ZIP, and **Type** for Factype_1. Click OK. Now headers for these attributes will show the aliases instead of original names.

3 Turn off all fields for Streets except Street and give it the alias **Name**.

4 Try using the Identify tool for Food Facilities and Streets to see the visible columns and aliases. The results are useful and clearly labeled. This is very helpful to the map user.

Set map tips

When you hover over a feature, you can get an attribute to display in a small window called a map tip.

1 Open the Display tab of Layer Properties for Food Facilities.

2 Select Address for the Display Expression and click Show Map Tips using the display expression. After you click OK, whenever you hover over a Food Facility, you will get its address in a map tip window.

3 Click OK and hover over some food facilities to see the map tips.

Create map hyperlinks

If there were an attribute with URL website addresses for food facilities, you could select it in the Display tab of Properties and turn mapped features into hyperlinks. That's not the case here, nevertheless, you can manually enter some hyperlinks to features, which you do next.

1 Zoom in to the area with the food pantry, West Side Campaign Against Hunger.

2   Click the Identify tool on the Tools toolbar.

3   Click the point marker for the West Side Campaign Against Hunger food pantry.

4   In the Identify window, right-click the address, 263 W 86 St, in the top panel and click Add Hyperlink.

5   Click Link to URL, type http://www.wscah.org/ in the corresponding textbox, click OK, and close the Identify tool.

6   Repeat steps 4 and 5 for the Goddard Riverside Community Ctr. and use http://goddard.org/ for the URL.

The remaining food facility in this group of three does not have a website as yet. So instead you can place a hyperlink to a document, in this case a PDF on your computer's hard drive.

7   With the Identify tool, click the point marker for the Westside Federation for Seniors and Supportive Housing, right-click its address, click Add Hyperlink. Click the browse button for Link to Document, browse to the Data\DataFiles folder, click Westside Federation.pdf > Open > OK. Close the Identify window.

8   Click the Hyperlink tool on the Tools toolbar ⚡ (all features with hyperlinks available get a blue dot in their point markers) and try out your hyperlinks. You have to get the tip of the hyperlink tool's lightning bolt exactly in the center of a blue dot so that the lightning bolt turns black. Then you can click to activate the link.

9   Save your map document.

Tutorial 3-2

Creating map layouts

In this tutorial you learn how to create map layouts in ArcMap for use in a Word document, PowerPoint presentation, or website — or for distribution as paper maps. Map layouts have several elements including a title, the map itself, a legend, a scale bar for ground distances, notes for data sources, and so forth. Rather than using pre-built layout templates, you learn to build your own layouts from scratch — an essential skill.

Open a map document and save a layer file

The set of layouts that you produce in this tutorial are for comparing populations by race or ethnicity for states. To facilitate comparisons of populations on separate maps, it's desirable to use the same numeric scale for all maps. Have you ever tried to compare bar charts in a newspaper or other publication where the horizontal-axis categories are the same but vertical axis scales are different? In your mind you have to attempt to transform them to the same scale in some way to make comparisons. While you might be able to better represent each map separately with its own custom break points, for comparison purposes it's better to use the same break points for all maps. So as part of the work, you save a layer file that allows easy reuse of a numeric scale.

1 Open Tutorial3-2.mxd from the Maps folder.

2 Save the map document as **Tutorial3-2Asians.mxd** to the Chapter3 folder.

3 Right-click Population in the table of contents, click Save As Layer File, browse to the Chapter3 folder of MyExercises, type **StatesPopulation.lyr** for Name, and click Save.

The image on the following page is the custom layout you will build, shown in design mode where you construct layouts. The blue horizontal and vertical lines are guides that you create for precise alignment and placement of elements. They don't show up on the layout when you print it or save it as an image.

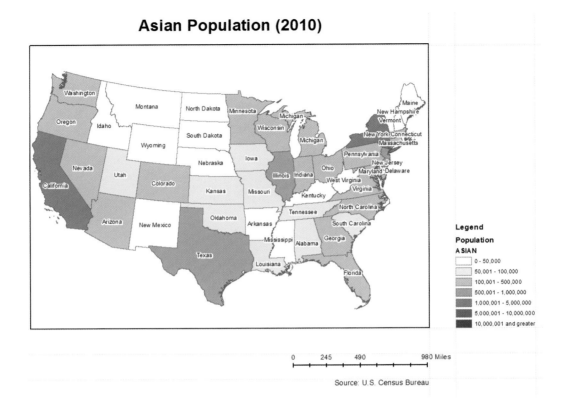

Verify Layout View options

You need to verify option settings for working on layouts. In particular, you need settings to create and show guidelines and have layout elements snap to them. Snapping makes alignment of layout elements easy and precise.

1 On the Menu bar, click View > Layout View.

2 On the Menu bar, click Customize > ArcMap Options > Layout View tab.

3 Verify that your settings match those with the image on the next page.

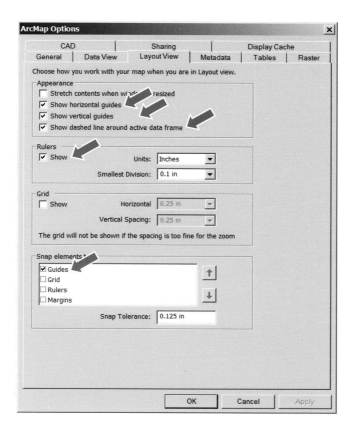

3-1
3-2
3-3
3-4
3-5
3-6
3-7
3-8
A3-1
A3-2
A3-3
A3-4

4 When finished, click OK.

Set up layout page orientation and size

Next, set up the layout page assuming that you will use an 8.5-by-11-inch document.

1 Right-click anywhere in the white space outside the layout and click Page and Print Setup.

2 If you have access to a printer, select it and desired properties, and then close any windows for setting printer properties.

3 In ArcMap's Page and Print Setup window, select Letter (8.5 by 11 inches) for paper size and Landscape for the Orientation in both the Paper and Page panels. See the image on the next page. Be sure to select the option buttons for landscape orientation in two places on the Page and Print Setup window.

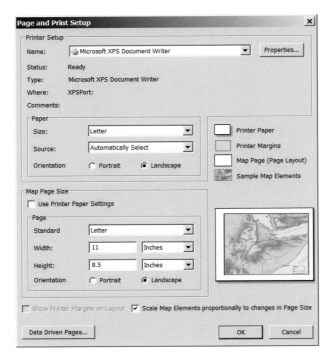

4 Click OK.

5 Right-click anywhere inside the layout and click the Zoom Whole Page button.

Create and use guidelines in the layout view

In the next steps, you will use vertical and horizontal rulers to create guidelines.

1 Click at 8.5 inches on the top horizontal ruler to create a vertical blue guide at that location. If you place your guide at the wrong location, right-click its arrow on the ruler, click Clear Guide, and start over. Alternatively, you can drag its arrow to relocate it.

2 Do the same at 7 inches on the left vertical ruler.

3 Click the map to select it (its frame outline and grab handles appear), right-click the map and click Properties, and then click the Size and Position tab.

4 Select the Preserve Aspect Ratio check box, type **7.5** in the Size Width field, press the TAB key, and click OK. Your map is now zoomed in too far, but its dimensions are good for the layout.

5 Drag the map so that its upper right corner is at the intersection of the two guides. The map snaps precisely to the intersection of the guides when you release.

The objective of the next step is to fill the map element rectangle with the map as large as possible.

6 Use the Zoom tools to maximize the map of the contiguous 48 states within its map element rectangle. If you need to start over, click the Full Extent button.

Insert a title

1 On the Menu bar, click Insert > Title and double-click the title.

2 Type **Asian Population (2010)** in the text box, replacing the code enclosed in angle brackets.

3 Click Change Symbol, select 22 for size, and click B (Bold) for Style.

4 Click OK > OK.

5 Center the title over the map.

Insert a legend

1 Click the horizontal ruler at 10.5 inches to create a new vertical guideline, and click the vertical ruler at 1.5 inches to create a new horizontal guideline.

2 Click Insert > Legend (Population should be in the right panel, Legend Items), click Next four times, and click Finish.

Resizing and placing the legend takes the next three steps.

3 Drag the legend so that its right side snaps to the 10.5-inch vertical guide and bottom to the 1.5-inch horizontal guide.

4 Click the horizontal ruler at 9 inches to create a new vertical ruler.

5 Drag the top left grab handle of the legend to the right and down to make the legend smaller. Snap it to the 9-inch vertical guideline on the left while staying locked to the 10.5-inch vertical guideline on the right and 1.5-inch horizontal guideline on the bottom.

Insert a scale bar

1 Click the vertical ruler at 1 inch to create a new horizontal guide.

2 Click Insert > Scale Bar.

3 Click Scale Line 2 > Properties.

4 Select Miles for the Division Units, click OK > OK.

5 Drag the scale bar so that its top is at the 1-inch horizontal guideline and its right side is at the 8.5-inch vertical guideline.

6 Right-click the scale bar and click Zoom to Selected Elements.

7 Drag the left side of the scale bar to the right until its width is 1,000 miles. This takes trial and error, with you dragging and releasing to see the resultant width in miles.

8 Right-click anywhere in the layout and click Zoom Whole Page.

Insert text

1 Click the vertical ruler at 0.5 inches to create a new horizontal guideline.

2 Click Insert, Text. ArcMap places a small text box in the center of your map (it is difficult to see at this scale).

3 Double-click the text box, type **Source: US Census Bureau**, and click OK.

4 Drag the text box so that its top right corner is at the intersection of the 8.5-inch vertical and 0.5-inch horizontal guides. That completes the layout.

5 Save your map document.

GIS TUTORIAL 1

GIS Outputs   CHAPTER 3   109

3-1
3-2
3-3
3-4
3-5
3-6
3-7
3-8
A3-1
A3-2
A3-3
A3-4

Create a JPEG image of the layout

1 Click File > Export Map.

2 Save the image with file name **Tutorial3-2Asians** to the Chapter3 folder of MyExercises with Save as type JPEG and resolution 300 dpi (publication quality, a big file).

3 Open a Computer window, browse to the Chapter3 folder of MyExercises and double-click Tutorial3-2Asians.jpg to open in a viewer. It is a very sharp and professional-appearing map.

4 Close the viewer and the Computer window.

Tutorial 3-3

Reusing a custom map layout

Reusing your custom map to produce additional maps saves time, but just as important is the consistency of the resulting maps. Reuse guarantees that sizes and placements of objects match perfectly for a collection of maps.

Open a map document

1 Open Tutorial3-3.mxd from the Maps folder.

You will replace the Asian attribute with the Black attribute of the Population layer.

2 Save the map document as **Tutorial3-3Blacks.mxd** to the Chapter3 folder.

Use a layer file

1 Click View > Data View.

2 Right-click the Population layer in the table of contents, click Properties > Symbology tab > the Import button.

3 In the Import Symbology dialog box, click the browse button. Browse to the Chapter3 folder of MyExercises, double-click StatesPopulation.lyr, and click OK.

4 In the Import Symbology Matching dialog box, click the Value Field arrow, click Black > OK.

5 In the Layer Properties window, change the color ramp to a monochromatic blue ramp, right-click the color ramp (to the right of the label, Color Ramp), click Properties > the arrow for Color 1 > the white color chip > button for Color 2 > its arrow > a dark blue > OK > OK.

6 Click View > Layout View.

7 Change the map title to **Black Population (2010)** and click OK.

Black Population (2010)

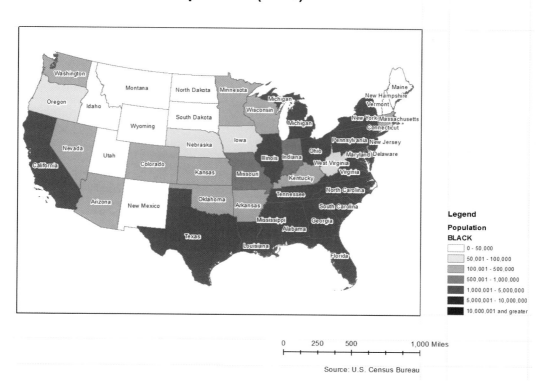

8   Save your map document.

3-1
3-2
3-3
3-4
3-5
3-6
3-7
3-8
A3-1
A3-2
A3-3
A3-4

YOUR TURN

Make progress on completing the map collection of population by race by making a map document for one of the following populations from your template: whites, Hispanics, or Native Americans. Name the map **Tutorial3-Whites.mxd**, **Tutorial3-Hispanics.mxd**, or **Tutorial3-NativeAmericans.mxd**, respectively, and save it to the Chapter3 folder of MyExercises. Use a monochromatic color ramp of your choice, but reuse the StatesPopulation.lyr layer file for break points.

Tutorial 3-4

Creating a custom map template with two maps

To facilitate comparisons, you can place two or more maps on the same layout. Your population maps by racial/ethnic groups are ideal for this purpose because they share the same numeric scale, making comparisons easy.

Open a map document

1 Open Tutorial3-4.mxd from the Maps folder and save the map document as **Tutorial 3-4AsiansBlacks.mxd** to the Chapter3 folder.

2 Use your zoom tools to get the contiguous 48 states to fill the map window, without Alaska showing.

3 Right-click the Population map layer and turn off Label Features for now. Having to wait for ArcMap to add labels every time you make a change to the data frame takes too long when you get to layout view. So to speed up interactions you turn labeling off. You can turn it back on after the layout is finished. Nevertheless, ArcMap may slow down in reacting to your changes in this tutorial, so be patient. Layout view is a heavy user of computer resources.

Create a second data frame

So far, you have worked with map documents that have only one data frame in the table of contents. It's possible to have two or more data frames, and this is necessary if you want to have two or more maps in a layout, as in this tutorial. While you could create a new data frame from scratch by clicking Insert > Data Frame, the easiest thing to do is to copy and paste the existing data frame, and then modify or add to its map layers. One of the benefits of copying and pasting is that the existing data frame has a map projection applied (Albers Equal Area Conic in this case), and when you copy and paste it to create a new data frame the projection is retained. Otherwise, if you were to create a new data frame from scratch, you'd have to add the projection as a separate series of steps.

1 In the table of contents, right-click Layers, click Properties > General tab, and change the name from Layers to **Asians** (but do not click OK).

2 Click the Coordinate System tab and notice that the projection is USA Contiguous Albers Equal Area Conic. This projection flattens the portion of the globe that includes the contiguous 48 states.

3 Click OK > Yes.

4 Right-click the Asians data frame and click Copy.

5 Click Edit > Paste. ArcMap adds a copy of the data frame to the table of contents.

6 Change the pasted data frame's name from Asians to **Blacks** and click OK > Yes.

3-1
3-2
3-3
3-4
3-5
3-6
3-7
3-8
A3-1
A3-2
A3-3
A3-4

YOUR TURN

For symbology of the Population map layer in the Blacks data frame, right-click Population, click Properties and the Symbology tab. Import the StatesPopulation.lyr layer file using Black as the value field. Keep the same color ramp as the Asians data frame. Both data frames then can use the same legend in the map layout. Your finished map document will appear as follows (shown with labels on, but which are off for you). Note that to switch from one data frame to another, you right-click the data frame and click Activate. The active data frame's label in the table of contents is in bold type.

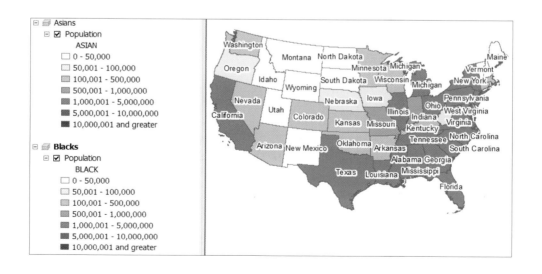

Create guidelines in layout view

At right is the layout that you will create in the steps of this exercise. It makes comparisons between two population distributions easy.

1   Click View > Layout View. Both data frames are in the view, but may be directly over each other so that you see only one.

2   Click the visible view, drag it up an inch or so, and release so that you see both views.

3   In layout view, right-click in the layout, click Page and Print Setup, make sure Size is Letter, make sure that both Portrait radio buttons are selected in the Page and Paper frames, and then click OK.

4   Click the horizontal ruler at the 0.5-, 6.5-, 6.8-, and 8.0-inch marks.

5   Click the vertical ruler at the 1.0-, 5.4-, 5.6-, 6.5-, and 10.0-inch marks.

Relocate and resize data frames

1 Make sure that the Asians data frame is active. Its data frame name needs to be bold in the table of contents and have grab handles visible on its boundary in the layout.

2 Drag the Asian data frame so that its upper left corner snaps to the intersection of the 10.0-inch horizontal guide and the 0.5-inch vertical guide.

3 Drag the lower right grab handle of the data frame to snap it at the 6.5-inch vertical guide. Drag a bottom-line grab handle to move the bottom of the data frame to the 5.6-inch horizontal guideline.

4 Drag the Blacks data frame so that its upper left corner snaps at the intersection of the horizontal 5.4-inch guide and the vertical 0.5-inch guide.

5 Drag the Blacks data frame so that its lower right grab handle snaps at the 6.5-inch vertical guide. Drag the bottom of the data frame to the 1.0 horizontal guide.

Change and coordinate map extents of both data frames

The objective of this exercise is to get the map of each data frame to fill its frame and be identical in extent in both frames. Unfortunately, there is not a direct way to accomplish this, but there is a good guide available called "extent indicators" that you use to get close to the objective.

1 Make sure that the Asians data frame is active. Use your Zoom and Pan tools to get the contiguous 48 states to fill up the data frame and be centered. Be careful that the Pan button does not get clicked on or remain on after use. It's easy to accidentally change the map view and extent with this tool. So always click the Select Elements tool on the Tools toolbar after using the Pan tool.

2 Make Blacks the active frame.

3 Right-click the Blacks data frame and click Properties > Extent Indicators.

4 Move Asians from the left panel to the right panel and click OK.

5 Click the Fixed Zoom Out button ⬚ on the Standard toolbar so that you see the entire red extent indicators.

3-1
3-2
3-3
3-4
3-5
3-6
3-7
3-8
A3-1
A3-2
A3-3
A3-4

6 Use the Zoom In tool to drag a rectangle matching the red rectangle of the extent indicators. You can repeat steps 4 and 5 until you get the Blacks data frame close to the extent of the Asians data frame.

7 Click the Select Elements button on the Tools toolbar. Right-click the Blacks data frame and click Properties > Extent Indicators. Move Asians from the right panel to the left panel and click OK. Now both frames should have maps that fill up frames and match well.

Insert a legend

1 Click the Asian map element in the layout to activate its frame.

2 Click Insert > Legend.

3 Click Next four times, and then click Finish.

4 Drag the legend so that it snaps on the upper right to the 8-inch vertical guide and 6.5-inch horizontal guide intersection, and then resize it to fit between the 6.8-inch and 8-inch vertical guides.

Convert the legend to a graphic and modify it

The legend has the heading word "ASIAN", which you need to delete so that the legend is general and applies to both data frames. You can edit the legend by turning it into a graphic and ungrouping its parts.

1 Right-click the legend and click Zoom to Selected Elements.

2 Right-click the legend and click Convert To Graphics. Now the legend is no longer a "live object" that is automatically updated if you resymbolize the Asian Population layer. It's a static graphic, but now you can edit it.

3 Right-click the legend and click Ungroup. While not needed here, you could further ungroup paint chips from their numeric intervals and edit the interval text.

4 Click anywhere outside the legend to deselect its parts, click the ASIAN textbox, and press DELETE. Drag the Legend and Population textboxes down to place them closer to the rest of the legend. An alternative that often works better than dragging is to select an element, such as the Population text box, and use your arrow keys (down arrow in this case) to move it.

5 Drag a rectangle around all legend parts to select them, right-click the legend and click Group. Now the parts won't get separated.

6 Right-click anywhere in the layout and click Zoom Whole Page.

Insert a title

1 Click Insert > Title. Click anywhere outside the title's text box, right-click the text box, and click Properties and the Text tab.

2 In the Text panel, type **Asian and Black**, press ENTER to jump to a new line, type **Populations (2010)**, type **90** for Angle, and press TAB. Click Change Symbol, change the Size to 20, click B (bold) for style, and click OK > OK.

3 Position the top left of the title text box at the intersection of the 6.8-inch vertical and 10-inch horizontal guides.

4 Save your map document.

YOUR TURN

Add a text box with the word "Asians" to the top map and another text box with "Blacks" to the bottom map. Export your layout to the Chapter3 folder of MyExercises as a JPEG image and view it in an image viewer. It is quite a nice layout and image. While there are clear differences, there are also some remarkable similarities in the distributions of the two races.

Tutorial 3-5

Adding a report to a layout

Sometimes it's helpful to add a tabular report to a layout. The easiest thing to do is to export the map data of interest to Excel, create the tabular display you want, and copy and paste it to your layout. Here you use an Excel worksheet that is already prepared with a table of the top 10 states by Asian population.

Open a map document

1 Open Tutorial3-5.mxd from the Maps folder and save the map document as **Tutorial 3-5AsiansReport.mxd** to the Chapter3 folder. This is already a completed layout, except there is no data table as yet.

Open an Excel workbook

1 Launch Microsoft Excel, click File > Open, browse to the Data > DataFiles folder, and open AsianTop10States.xlsx. The worksheet has a nice table created from the population map layer's attribute table. The table was exported when it was open in ArcMap using the Table Options button and extra columns and rows were deleted, leaving the state names and populations of the 10 states with the highest Asian population. Features of the worksheet include that the grid is turned off (View tab, Show panel, Gridlines unchecked), borders are included for data cells (Home tab, Font panel, All Borders), and cells for the range A1:B12 (columns A and B and rows 1 through 12) have a white fill color (Home tab, Font panel, Fill Color, white).

	A	B
1	**Top 10 States**	
2	**State**	**Asian Population**
3	California	4,861,007
4	New York	1,420,244
5	Texas	964,596
6	New Jersey	725,726
7	Illinois	586,934
8	Hawaii	525,078
9	Washington	481,067
10	Florida	454,821
11	Virginia	439,890
12	Massachusetts	349,768

2 Select the range A1:C13. You have to include an extra column and row in order to get all borders to copy. The extra column (C) and extra row (13) do not have color fill, so they will be invisible when pasted into the layout.

3 Press CTRL+C to copy the range.

4 Click the layout in ArcMap.

5 Click Edit > Paste.

6 Relocate and resize the table as seen in the following:

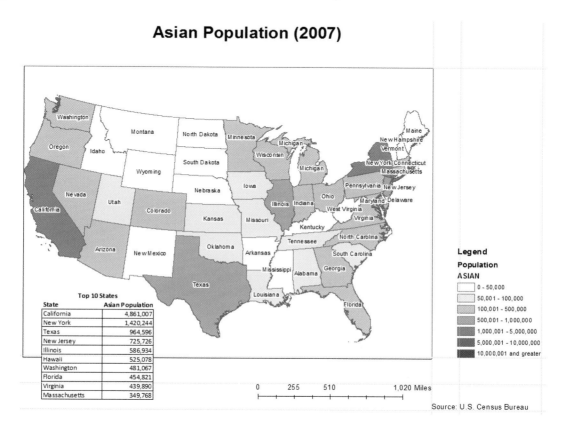

7 Save your map document.

Tutorial 3-6

Adding a graph to a layout

ArcMap has a convenient and easy means to create and add graphs to a layout.

Open a map document

1 Open Tutorial3-6.mxd from the Maps folder. This is the same data view with which you started the previous tutorial.

2 Save the map document as **Tutorial3-6AsiansGraph.mxd** to the Chapter3 folder.

Sort and select records

1 Right-click Population in the table of contents and click Open Attribute table.

2 Right-click the ASIAN header and click Sort Descending.

3 Select the first 10 records (California through Massachusetts) by clicking and dragging the row selectors for those rows.

4 Close the attribute table.

Create a graph and add it to a layout

1 Click View > Graphs > Create Graph. Then make selections as shown:

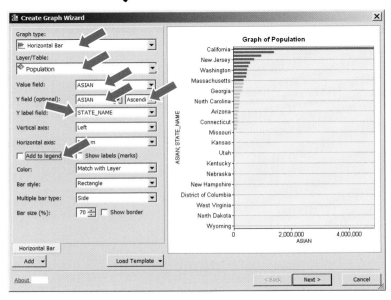

2   Click Next and make selections as follows:

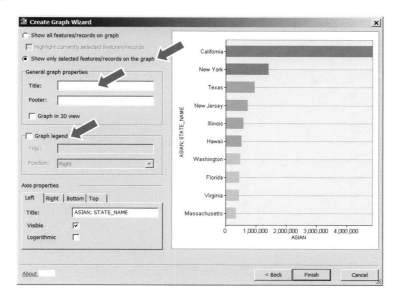

3   Click Finish.

4   Right-click the graph, click Save, browse to Chapter3 of the MyExercises folder, type **AsianTop10Graph**, and click Save.

5   Right-click the graph, click Add to Layout, and close the graph window.

6   Snap the lower left corner of the graph to the intersection of the 0.5-inch vertical and 0.5-inch horizontal guides.

7   Save your map document.

Tutorial 3-7

Building a map animation

Police want to identify new, persisting, and fading spatial clusters of crime locations. These clusters make good areas for police to patrol for enforcement and prevention. Animations of crime data for this purpose are effective if they display two tracks of points: new crime points along with recent, but older crime points for context. Then the observer can detect the emergence of new clusters (where there were none before), the persistence of existing clusters getting new crime points, and fading clusters as no new points are added and the cluster fades away. So the approach you take calls for two animation layers—one for each day's events and a second with the past two weeks' events. Computer-aided dispatch (CAD) drug calls and shots-fired calls are important crime events for this kind of analysis.

Open the map document for animation

First you build an animation showing just the sequence of one set of points in a track, one day at a time. About all this accomplishes is to get you started with animation and convince you that the drug and shots-fired calls jumped around in a portion of the Middle Hill neighborhood of Pittsburgh. Then you add a second set of points that provide context, allowing you to better see hot-spot patterns.

1 Open Tutorial3-7.mxd from the Maps folder. The map document opens to the Middle Hill neighborhood. The CAD data, with dates ranging from 7/1/2012 through 8/31/2012, represents calls from citizens reporting illegal drug dealing and shots fired. The animation that you will build will show the daily sequence of CAD call locations, starting with 7/1/2012.

2 Save the map document to the Chapter3 folder.

Set time properties of a layer

1 Right-click CAD Calls in the table of contents, click Properties > Time tab, and type or make selections as follows (but do not click OK). CALLDATE has dates such as 7/1/2012. The Time Step Interval is the unit of time for measurement, here 1 day.

Layer Properties screen with arrows pointing to: "Enable time on this layer" checkbox, the CALLDATE Time Field, the Time Step Interval value "1" and "Days", the Time Zone dropdown, and the "Values are adjusted for daylight savings" checkbox.

2 Click Calculate, change the Time Step Interval from 4 to 1, and click OK.

Use the Time Slider window for viewing

With time properties set, you are ready to use the Time Slider interface to play a simple video of daily crime points.

1 On the Tools toolbar, click the Open Time Slider window button 🕐 .

2 In the Slider window, click the Options button ⋮≣ > Playback tab.

3 Drag the speed selector to roughly half way between slower and faster and click OK.

4 Click Enable time on the map button in the upper left corner of the Time Slider window.

5 Click the Play button ▶ on the slider.

The video plays, one day at a time, shown in the following image at July 30, 2009. Play the video a few more times to see if you can spot any patterns. This is hardly possible until you add the crime time context to the video.

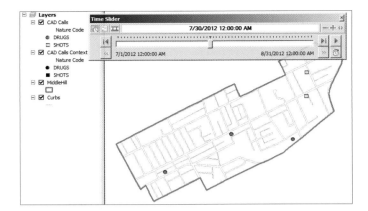

6   Close the Slider window and save your map document.

Create a new date column for animating a window of crime points

On any day of the animation, you need to show the current day's crimes with point markers in bright colors and the crime context consisting of two weeks' crime points ending on the same day with black point markers. Displaying a moving window of crime points for two weeks' data requires starting and ending dates for the window. For this purpose you create a new date column with 14 days added to CALLDATE to yield the end date.

1   Right-click CAD Calls Context (the layer with the black point markers) in the table of contents and click Open Attribute Table.

2   Click the Table Options button, select Add Field, type **EndDate** for Name, select Date for Type, and click OK.

3   Right-click the header for EndDate and click Field Calculator.

4   Double-click CALLDATE in the Fields panel, click the + button, and type a **blank space** and **14** to yield **[CALLDATE] + 14** for EndDate's expression.

5   Click OK. The value of EndDate for the first record is 8/3/2012, 14 days after the record's CALLDATE value of 7/20/2012.

6   Close the attribute table.

Set advanced time properties of a layer

1   From the table of contents, right-click CAD Calls Context, click Properties > Time tab, and type or make selections match those with the image on the next page.

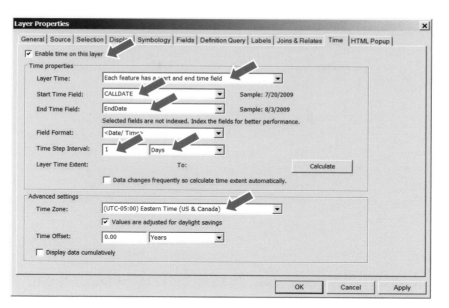

2 Click Calculate, change the Time Step Interval from 4 to 1, and click OK.

When you play the animation next, both layers with time properties will animate. CAD Calls will show the current day's crime locations while CAD Calls Context will show all crimes in the interval, including two weeks ending on the current day.

Use the Time Slider window for advanced viewing

1 On the Tools toolbar, click the Open Time Slider Window button.

2 Turn on the CAD Calls Context layer. Click the Play button on the slider.

Shown is the animation at July 30, 2012. You see all of the crime locations for the last two weeks with the last day's crimes in red or yellow and older crimes in black. At this time there is much persistence in crime clusters.

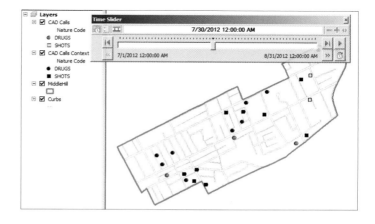

3 Save your map document.

3-1
3-2
3-3
3-4
3-5
3-6
3-7
3-8
A3-1
A3-2
A3-3
A3-4

Tutorial 3-8

Using ArcGIS Online

ArcGIS Online (www.arcgis.com) *can be used for storing, sharing, and using map layers in a browser. It provides 2 GB of free storage for your online maps and it provides access to online Esri- and user-supplied map layers. This allows you to prepare spatial data using ArcGIS for Desktop and then share it over the Internet using ArcGIS Online as your free web service host. As a preview, the following are the overall steps you have to take to author maps and publish them on the Internet as web services:*

- Sign up for an Esri Global Account. It's free and it gives you access to many ArcGIS resources.

- Convert map layers authored on ArcGIS for Desktop to shapefiles and compress them for upload to ArcGIS Online.

- Create and save online map compositions, based on spatial data from your desktop and/ or ArcGIS.com account, for the public or restricted groups to use.

Create an Esri Global Account

To use ArcGIS Online, create an Esri Global Account, which provides access to many additional resources available from www.esri.com.

1 In your browser, go to www.arcgis.com.

2 Click the Sign In link at the top right of the window.

3 Under "Don't have an Esri Global Account?" click Create a Personal Account.

4 Fill out the resulting form and click Create My Account. If you are in a class with an instructor, use the naming convention that your instructor provides, such as GISClassYourName. Make a note of your username and password.

5 Click ArcGIS on the Main Menu. This takes you to the ArcGIS Online home page, and you are signed in.

Let's pause here a moment to get an overview of the ArcGIS Online main menu:

- ArcGIS brings you back to the ArcGIS Online home page and access to the full menu of options.

- Gallery provides a collection of finished ArcGIS Online maps for you to use.

- Map provides the interface for creating your own online map compositions.

- Groups allows you to create, join, and activate groups.

- My Content provides utilities to maintain your map layers and compositions and layers.

- The search box allows you to search for map layers and compositions

Add desktop map data to a map in ArcGIS Online

You can use spatial data that you've prepared with ArcGIS for Desktop in your ArcGIS Online web maps. The two primary formats for vector map layers that ArcGIS Online accepts for upload are compressed (zipped) shapefiles for any kind of vector map layers and comma separated value (CSV) text files for points, that include latitude and longitude coordinates (XY data). Compressed shapefiles have the following requirements and limitations:

- The files making up the shapefile must be at the root of the compressed file and not in a folder. So in Computer, select the files making up the shapefile, right-click the selection, click Send to, and select Compressed (zipped) folder.

- At most there can be 1,000 features per shapefile.

To save time, spatial datasets are available in the EsriPress Data\DataFiles folder that you will use as local spatial data to upload to ArcGIS Online:

- PittsburghSchools.csv is a CSV text file for Pittsburgh public schools. Attributes include the following:
  - Name = school name
  - Level = type of school (primary, middle, or high)
  - Address
  - ZIPCode
  - Enroll = total enrollment
  - EnrollAsia = Asian student enrollment (likewise there is a breakdown for Hispanics, blacks, whites, and others)
  - Latitude
  - Longitude

• Neighborhoods.zip is a zipped shapefile for Pittsburgh neighborhoods that includes Hood as an attribute with neighborhood name.

1 In ArcGIS.com, click Map on the Main bar. A new map opens with a basemap displayed for the continental US.

2 Click the Add button arrow ✦ Add ▼ , click Add Layer from File, click browse, browse to \EsriPress\GIST1\Data\DataFiles\, click PittsburghSchools.csv file > Open > Import Layer. ArcGIS Online imports the file, zooms to Pittsburgh, displays a basemap, and displays the Pittsburgh schools layer.

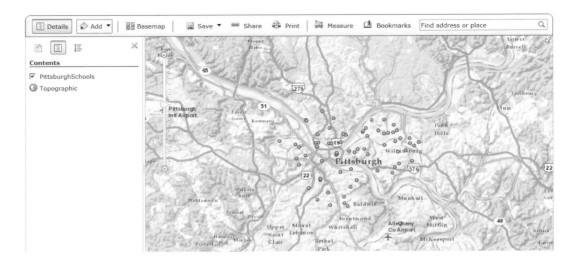

3 Click Save > Save As on the Main bar and enter text as follows. Replace "Jones, Bill" with your name or user name.

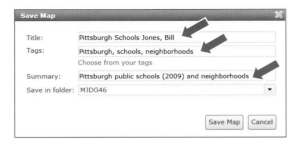

4 Click Save Map.

> **YOUR TURN**
>
> Add Neighborhoods.zip to your map using the same procedure. The imported neighborhoods cover up the schools, so hover over PittsburghSchools in the table of contents, click the small arrow that appears to its right, and click Move up. Save your map.

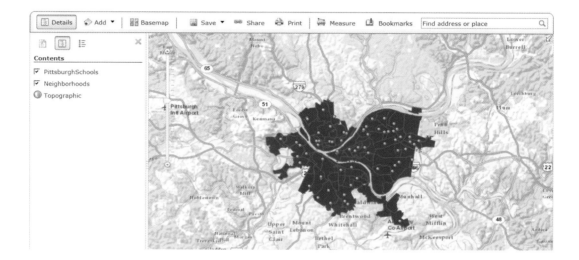

Symbolize map layers

ArcGIS Online gives you much of the same functionality to symbolize map layers as is available in ArcGIS for Desktop. Next, you symbolize the two layers that you added to your web map.

Rename layers

1 Hover over PittsburghSchools in the table of contents, click the options arrow that appears, click Rename, type **Pittsburgh Public Schools**, and click OK.

2 Do the same for neighborhoods, renaming it **Pittsburgh Neighborhoods**.

Symbolize polygons

1 Hover over Pittsburgh Neighborhoods in the table of contents, click the options arrow that appears, and click Change Symbols.

2 The default of A Single Symbol is what's needed here, but click the arrow to the right and look at the options. Single Symbol uses the same color fill and border for each polygon, Unique Symbols uses a code in the layer's attribute table for color coding, and Color applies a color ramp to a numeric scale that you design.

3 With A Single Symbol selected, click Change Symbol.

Needed here is hollow fill, which you can apply using the "trick" in the next step.

4 In the Transparency row, move the slider to 100%.

5 Type **2** for Outline pixel width (px).

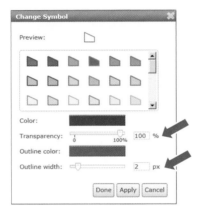

6 Close the Change Symbols window. Click Done.

7 Save your map.

Symbolize points

1 Hover over Pittsburgh Public Schools in the table of contents, click the options arrow that appears, click Change Symbols.

2 Select Unique Symbols, select the attribute, Level, to show.

3 Click the point marker symbol for High (high school) shown above and click Change Symbol.

4 Select Shapes for the point marker family (drop-down list right after Preview), select the blue circular point marker, and type **24** for the size (px).

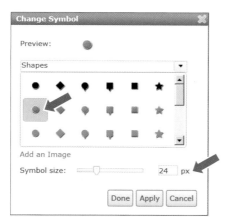

5 Click Done > OK > Apply.

6 Save your map.

YOUR TURN

Symbolize the remaining values of the Level code as you wish. Zoom farther in so that Pittsburgh fills more of the window. Close the Change Symbols window and click the Show Map Legend button ▤ . Try clicking a point to see that a window pops up with the school's record. It's possible to configure the pop-up, but you won't do that here.

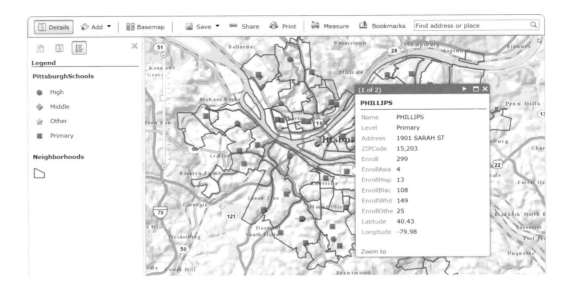

Add layers from ArcGIS Online

ArcGIS Online is not only a website for sharing maps that you create, it is also Esri's online repository of spatial data. Esri has maps available in ArcGIS Online from its own sources, its partners, and the GIS community at large. Anyone can share maps and data via ArcGIS Online. Next, you change the basemap and then search for hospitals to add to your map.

When you search for a layer using keywords, ArcGIS produces results that match the keywords and overlap with the map extent of your current window. Besides adding online spatial data, you can easily change the basemap, which is what you do next.

Change the basemap

1 Click the Basemap button 🔲 Basemap on the Main Menu.

2 Click Streets. ArcGIS replaces the former basemap with the Streets basemap.

3 Zoom farther in to Pittsburgh. More detailed streets turn on as you zoom in.

YOUR TURN

Experiment with other basemaps. Be sure to try some imagery and zoom way in. Save your map.

Search for and add a layer from ArcGIS Online

1 Click the Add button on the Main menu and click Search for Layers.

2 Type **hospitals** in the Find box, make sure that the In box is set to ArcGIS Online, and click Go.

3 Click Add for HHS Health Resources. That adds the layer to your map. If you hover over this layer in the legend and click the arrow that appears, you can click Show Item Details to view descriptive information about the layer.

4 Close the Search for Layers window, click Details, and Show Map Legend. Now you can interpret the added layer.

5 Save your map. Your map is finished.

Check the status of your map

Your map is accessible just by you at this point. However, if you have an instructor, he/she can create a group. Your instructor would click Groups > Create a Group, fill out a form, select Public, and click Users can apply to join group. Then you can request to join the group. Your instructor would check for membership requests to accept them by opening the group and clicking the Membership Request button. After you are accepted into the group, you can publish your map with the group and access any of the group's maps. Anyone in the group can access your shared maps. Alternatively, outside of the classroom and learning context, you could create your own group or join other existing groups.

1 Click My Content > My Content on the Main tool bar. You can see that ArcGIS has stored your map, Pittsburgh Public Schools Your Name, in a folder with your username. You can create new folders and store your work in new folders. Notice that your map is not shared.

2 Select the check box to the left of Pittsburgh Public Schools. Notice that this enables a couple of actions for the layer, including sharing it and deleting it.

3 Click the Share button. Your options for sharing are to share with everyone or with a group. Do not share your map with everyone (public) because that would just clog up the online system.

If your instructor has created a group, the following steps show you how to join the group and share your maps with it. Do nothing for now.

4 Click Cancel.

3-1
3-2
3-3
3-4
3-5
3-6
3-7
3-8
A3-1
A3-2
A3-3
A3-4

Search for and join a group

If your instructor did not create a group or if you are a self-learner, just read along. You won't be able to join a group.

1 Click Groups on the Main tool bar.

2 If your instructor has created a group, type the name of the group in the Search for groups textbox on the Main menu and press ENTER.

3 Click the name of the found group and click the Join This Group button.

4 Click Submit Request. You have to wait until your instructor grants your request. Then you are a member.

Share your maps with the group

Suppose that time has passed and your request to join the group is granted. When you click Groups, you will see that you are a member of the group.

1 Click MyContent on the Main toolbar.

2 Select the check box to the left of Pittsburgh Public Schools.

3 Click the line with Pittsburgh Public Schools (which is a hyperlink).

4 Click the Edit button and change the name of the map from Pittsburgh Public Schools to **Pittsburgh Public Schools by Your Name** where you substitute your name. Otherwise, others would have the same map title as yours so you need to add your name to differentiate your map in the group.

5 Click Save > Share.

6 Click the name of the group you are joining > OK. Now members of your group can use your map.

Using your maps on your smartphone or iPad

Free ArcGIS apps are available from Google Play, the Apple App Store, Amazon Appstore, and Windows Marketplace. Here is how to use the ArcGIS app on the iPhone:

1 Search for ArcGIS at the Apple App Store and install Esri ArcGIS on your iPhone.

2 Open Esri ArcGIS.

3 Click the Sign In button.

4 Enter your ArcGIS Online (Esri Global account) username and password.

5 Click My Maps.

6 Click your Pittsburgh Public Schools map. The map opens on your iPhone. You can use usual gestures to zoom in or out and pan.

7 Click the i button on the lower left of the screen. This provides access to three buttons: Legend for the map legend, Content for the table of contents, and Detail for documentation.

8 Click the Map button at the top left of the screen. You get back to your map.

3-1
3-2
3-3
3-4
3-5
3-6
3-7
3-8
A3-1
A3-2
A3-3
A3-4

Assignment 3-1

Create a dynamic map of historic buildings in downtown Pittsburgh

Walking tours are great attractions for tourists, as are historic sites and buildings. Here you create a dynamic map document—using visible scale ranges, hyperlinks, and map tips—for the Central Business District of Pittsburgh and some of its historic buildings.

The ArcMap document that you build can be considered a prototype for an ArcGIS Online map, available on the Internet.

While not a part of this assignment, you could export all map layers to shapefile format, compress each shapefile, and upload them to ArcGIS Online to rebuild the map there. For your information, to create visible scale ranges in ArcGIS Online, click the Show Contents of Map button, hover over a layer in the left panel, click the resulting arrow, and click Set Visibility Range. You create hyperlinks in an ArcGIS Online map using pop-up windows. Click the Show Contents of Map button, hover over a layer, click the resulting arrow, and click Configure Pop-up. Then under Pop-up Media, click Add > Image.

Get set up

- Rename the folder \EsriPress\GIST1\MyAssignments\Chapter3\Assignment3-1YourName\ to your name or student ID. Store all files that you produce for this assignment in this folder.
- Create a new map document called **Assignment3-1YourName.mxd** with relative paths.

Build the map

Add the following to your map document:

- \EsriPress\GIST1\Data\Pittsburgh\CBD.gdb\Outline—polygon feature of Pittsburgh's Central Business District neighborhood outline.
- \EsriPress\GIST1\Data\Pittsburgh\CBD.gdb\Bldgs—polygon features of Pittsburgh's Central Business District buildings. Attributes include: Name = name of the building (historic buildings only); Historic = 1 if an historic building, 0 otherwise; Hyperlinks = hyperlinks to websites for historic buildings only; and Address (historic buildings only).
- \EsriPress\GIST1\Data\Pittsburgh\CBD.gdb\Streets2—line features of CBD streets.
- \EsriPress\GIST1\Data\Pittsburgh\CBD.gdb\Curbs—line features of CBD curbs.
- \EsriPress\GIST1\Data\Pittsburgh\CBD.gdb\Histsites—polygon features of historic areas in Pittsburgh's Central Business District.

Requirements

- Use good labels with halos for streets, historic districts, and historic buildings.
- Symbolize curbs with a medium to dark gray and display them at all scales.
- Give streets no color (so that they are invisible), but label them. Make the labels visible when zoomed in to the largest historic area.
- Label historic areas at all scales
- Include two copies of buildings in the table of contents. Use a definition query to make one copy be historic buildings and the other nonhistoric buildings. Symbolize buildings so that nonhistoric buildings are in ground and historic buildings are figure (use a brown color fill).
- Display the historic buildings at all scales and nonhistoric buildings only when zoomed in to the largest historic area. Label historic buildings with Name when zoomed in to the largest historic district level.
- Use hyperlinks for buildings. *Hint:* Use the Display tab of the property sheet and the Hyperlinks panel there to turn on hyperlinks.
- Use Address for historic building map tips.
- Turn off all useless fields for all map layers. Add aliases if necessary.

3-1
3-2
3-3
3-4
3-5
3-6
3-7
3-8
A3-1
A3-2
A3-3
A3-4

Assignment 3-2

Create a layout comparing 2010 elderly and youth population compositions in Orange County, California

In this assignment, you create a map layout with two maps, one with population fraction of the elderly and the other with population fraction of youths in Orange County, California. The pattern of interest to show is that the fraction of youth population has a higher mean and variance than the fraction of elderly population.

Get set up

- Rename the folder \EsriPress\GIST1\MyAssignments\Chapter3\Assignment3-2YourName\ to your name or student ID. Store all files that you produce for this assignment in this folder.
- Create a new map document called **Assignment3-2YourName.mxd** with relative paths.

Create a map layout

Add the following to your map document:

- \EsriPress\GIST1\Data\UnitedStates.gdb\CAOrangeCountyTracts—polygon features for Orange County, California census tracts, Census 2010 with selected census variables: PopTot = total population, Pop0To17 = population ages 17 and younger, and Pop65Up = population 65 and older.

Requirements

- Include an 8.5-by-11-inch portrait layout with two data frames: one with the fraction of 2010 population 0–17 years old and the second with the fraction 65 or older. **Hint:** Use the Symbology tab in Layer Properties to show a population as a fraction of the total population. Use PopTot field to normalize the data.
- Use the same numeric and color scale for both maps. Use one legend entry for the two choropleth maps.
- Make both data frames the same size. You need to make both data frames in the layout identical in size and alignment. **Hint:** After approximately aligning and sizing the two frames in the layout, right-click a data frame's name in the table of contents and click Properties > Size and Position. Change the frame's width and height to "nice" numbers, for example from 4.9781 to 5.0, and make note of them. Then change the other data frame's width and height to the same values. Use a vertical guide line in the layout to align the two frames on the left.

- Zoom in to the same populous northwestern half of the county in both frames. ***Hint:*** Use the approach in tutorial 3-4 with extent indicators to get both frames to have the same extent.
- Include a title, graphic scale bar in miles, and your name as map author.
- Export the map as a JPEG file called **Assignment3-1YourName.jpg**.

3-1
3-2
3-3
3-4
3-5
3-6
3-7
3-8
A3-1
A3-2
A3-3
A3-4

Assignment 3-3

Create an animation for an auto theft crime time series

Auto thieves are often creatures of habit; they return to the same areas and repeat patterns that led to successful thefts in the past. An animation of successive auto theft locations can thus help determine the space-time pattern of such a thief. Suppose that police suspect a serial auto thief who steals cars for basic transportation and then vandalizes abandoned stolen cars in a unique way with spray paint.

Get set up

- Rename the folder \EsriPress\GIST1\MyAssignments\Chapter3\Assignment3-3YourName\ to your name or student ID. Store all files that you produce for this assignment in this folder.
- Create a new map document called **Assignment3-3YourName.mxd** with relative paths.

Create an animation of serial auto thefts

Add the following to your map document:

- \EsriPress\GIST1\Data\DataFiles\AutoTheftCrimeSeries.shp—point features of the suspected crime series of auto thefts.
- \EsriPress\GIST1\Data\Pittsburgh\MidHill.gdb\Outline—polygon feature of the Middle Hill neighborhood boundary.
- \EsriPress\GIST1\Data\Pittsburgh\MidHill.gdb\Streets—line features of streets in the Middle Hill neighborhood.
- \EsriPress\GIST1\Data\Pittsburgh\MidHill.gdb\Curbs—line features of curbs in the Middle Hill neighborhood.

Requirements

- Use a thick outline for the study area outline, use no color for Streets (so that they are invisible) but label them with a medium-gray font and use the same medium gray for Curbs. Add two copies of AutoTheftCrimeSeries.shp, one called Auto Theft Crime Series with a size 10 Circle 2 point marker and bright color fill, and the second called Auto Theft Crime Series Context with the same point marker but black color fill and placed under the first copy in the table of contents.

- Set the time properties for the two copies of AutoTheftCrimeSeries similar to those in tutorial 3-7. Have Auto Theft Crime Series display a single day's auto thefts in each frame and have Auto Theft Crime Series Context display the cumulative set of crimes. Set the playback speed to a medium level.

- Use the Export to Video button on the Time Slider window to create a movie file, **Assignment3-3YourName.avi**. Try playing the movie by double-clicking it in a Computer window.

3-1
3-2
3-3
3-4
3-5
3-6
3-7
3-8
A3-1
A3-2
A3-3
A3-4

Assignment 3-4

Create a shared map on ArcGIS Online

Add a map layer from your desktop computer to ArcGIS Online, symbolize it, add two layers of your choice from ArcGIS Online, and share the results with the group that your instructor creates. If you are not in a class, you don't have to share your map with anyone.

Start with the following:

\EsriPress\GIST1\Data\DataFiles\DCTract.shp—seven files making up the shapefile for the polygon features of Washington, D.C.'s census tracts. Included are selected 2010 census variables:

- PerCapInc = mean per capita income ($)
- PopTot = total population
- PopTotDen = total population density (persons/sq. mile)
- Pop0to17 = population under 18 years of age
- Pop65_ = population age 65 or older
- Pop18To24 = population age 18 to 24
- PopLTHighS = population age 18 to 24 with less than high school education
- PopBachDeg = population age 18 to 24 with a bachelor's degree or higher
- Workers = number of persons working in civilian jobs
- WorkerPub = number of workers in civilian jobs who take public transportation (except for taxi cabs) to work
- PWorkerPub = percentage of workers in civilian jobs who take public transportation (except for taxi cabs) to work

Create an ArcGIS Online map

Requirements are:

- Copy the seven files of the DCTract shapefile to a location you choose on your local computer. Compress all seven files to produce DCTracts.zip. **Hint:** Select all seven files of the shapefile in a Computer window, right-click the selection, and click Send to > Compressed (zipped) folder.
- Add the zipped shapefile to your online map and symbolize it for one of the available census variables using a color ramp and numeric scale of your choice.
- In ArcGIS Online, find two online layers relevant in some way for your census tract variable, and add them to your map.
- Save your map with the name **Assignment3-4YourName**.
- Share your finished map with the group your instructor creates, or if you don't have an instructor simply do not share your map document.

Online map layers that you add to your map will be below your census-variable, choropleth map in the table of contents. ArcGIS Online won't let you move online map layers up in the table of contents, and the order in which you add layers doesn't help—the online map layers always go on the bottom. So your choropleth map will cover the online map layers. The only available remedy is to make your choropleth map partially transparent (option when you hover over the table of contents entry for your choropleth map in Show Contents of Map mode).

3-1
3-2
3-3
3-4
3-5
3-6
3-7
3-8
A3-1
A3-2
A3-3
A3-4

File geodatabases

ArcGIS can directly use or import most GIS file formats in common use for geo-processing and display. The recommended native file format for use in ArcGIS for Desktop is the file geodatabase. It stores map layers, data tables, and other GIS file types in a system folder that has the suffix extension .gdb in its name. In this chapter you learn about working with file geodatabases.

Learning objectives

- *Build file geodatabases*
- *Use ArcCatalog utilities*
- *Modify attribute tables*
- *Join tables*
- *Aggregate data*

Tutorial 4-1

Building a file geodatabase

A file geodatabase is quite simple and flexible, being merely a collection of files in a file folder. Nevertheless, a file geodatabase provides a powerful structure and organization for spatial data. You need a special utility program to build and maintain a file geodatabase, ArcCatalog, which you use next. Some of the functionality of ArcCatalog is also available in ArcMap's Catalog window. The Catalog window allows you to do most data utility work while in ArcMap without opening the separate application program, ArcCatalog.

Open ArcCatalog

1 On the Windows taskbar, click Start > All Programs > ArcGIS > ArcCatalog 10.1.

2 Click the Connect To Folder button 🗁 on the Standard toolbar, expand the folder and file tree under Computer for EsriPress, click the GIST1 folder icon to select it, and click OK. The Connect To Folder button is important to remember. It sets roots for the folder and file tree that you use in ArcCatalog, providing access to files.

Create a new file geodatabase

While file geodatabases are only folders, you must create them using ArcCatalog or Catalog. Windows Explorer or Computer, while capable of creating a folder, cannot build a file geodatabase.

1 In the Catalog Tree panel, expand Folder Connections, EsriPress > GIST1, and the MyExercises folder.

2 Click the Chapter4 folder to display its contents (now empty) in ArcCatalog's right panel.

3   Right-click Chapter4 in the left panel and click New > File Geodatabase.

4   Change the name from New File Geodatabase.gdb to **Chapter4.gdb**. ArcCatalog creates a file geodatabase that you can populate with feature classes, stand-alone tables, and other objects. Feature classes are map layers stored in a geodatabase.

Next, you import map layers in shapefile format into your new file geodatabase as feature classes.

Import shapefiles

A shapefile is an older Esri file format that many spatial data suppliers still use for distributing map layers. ArcCatalog allows you to import shapefiles and other map layer formats into a file geodatabase.

1   In the ArcCatalog's right panel, right-click Chapter4.gdb, click Import > Feature Class (multiple). The multiple import option provides the convenience of importing several features at the same time as well as automatically naming output feature classes with the shapefiles' names.

2   In the Feature Class to Geodatabase (multiple) dialog box, click the browse button to the right of the Input Features field, browse to Data > MaricopaCounty, double-click to open that folder, hold the SHIFT key down, and select tl_2010_04013_cousub10.shp and tl_2010_04013_tract10.shp.

3   Click Add. That action adds tl_2010_04013_cousub10.shp and tl_2010_04013_tract10.shp to the input panel.

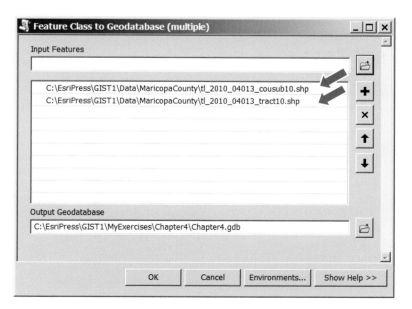

4 Click OK. ArcCatalog imports the shapefiles into the file geodatabase.

5 In the Catalog tree, expand the Chapter4 folder and Chapter4.gdb to see the imported feature classes.

Import a data table

Next, you import a data table with some 2010 census data at the tract level for Maricopa County, Arizona.

1 Right-click the Chapter4 file geodatabase, then click Import > Table (single).

2 In the Table to Table dialog box, click the browse button to the right of Input Rows, browse to the Data > MaricopaCounty folder, click CensusData.xlsx > Add > CensusData$ > Add.

3 Type **CensusData** in the Output Table field. Notice the Field Map panel listing the data fields that are imported.

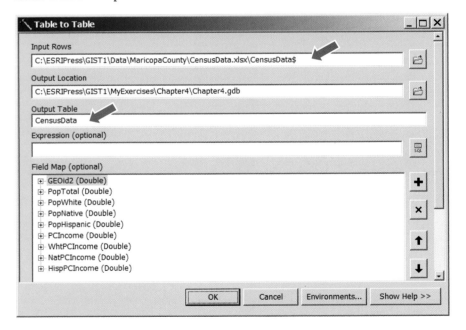

4 Click OK.

Tutorial 4-2

Using ArcCatalog utilities

Now that you've created a file geodatabase, you can start using ArcCatalog's utilities. First is Preview, which gives you a good overview of a feature layer or table.

Preview layers

1 If necessary, click Chapter4.gdb in the Catalog tree to expose its contents in the right panel.

Contents	Preview	Description	

Name	Type
CensusData	File Geodatabase Table
tl_2010_04013_cousub10	File Geodatabase Feature Class
tl_2010_04013_tract10	File Geodatabase Feature Class

2 In the right panel, click tl_2010_04013_cousub10 and click the Preview tab. ArcCatalog previews the map layer's geography.

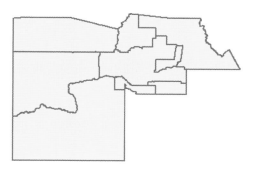

3 At the bottom of the Preview window, select Table as the Preview. ArcCatalog previews the map layer's attribute table.

OBJECTID *	Shape *	STATEFP10	GEOID10	NAME10
1	Polygon	04	0401390459	Buckeye
2	Polygon	04	0401390561	Chandler
3	Polygon	04	0401390867	Deer Valley
4	Polygon	04	0401391377	Gila Bend
5	Polygon	04	0401392601	Phoenix
6	Polygon	04	0401393009	St. Johns
7	Polygon	04	0401393060	Salt River
8	Polygon	04	0401393472	Tonto National Forest
9	Polygon	04	0401393774	Wickenburg

4 Click the Description tab. ArcCatalog previews the map layer's metadata in a report format. For TIGER maps, only general metadata is available.

Description
The TIGER/Line Files are shapefiles and related database files (.dbf) that are an extract of selected geographic and cartographic information from the U.S. Census Bureau's Master Address File / Topologically Integrated Geographic Encoding and Referencing (MAF/TIGER) Database (MTDB). The MTDB represents a seamless national file with no overlaps or gaps between parts, however, each TIGER/Line File is designed to stand alone as an independent data set, or they can be combined to cover the entire nation.

5 Click the Contents tab.

YOUR TURN

Preview tl_2010_04013_tract10 and CensusData.

Rename feature layers

Because a file geodatabase has a special file format, you must use ArcCatalog for file-management purposes, including renaming and copying items.

1 In the left panel under Chapter4.gdb, right-click tl_2010_04013_cousub10, click Rename, type **Cities**, and press the TAB key.

2 Repeat step 1 to rename CensusData to be **CensusTractData**, and tl_2010_04013_tract10 to be **Tracts**.

Copy and delete feature layers

1 In the left panel under Chapter4.gdb, right-click Cities, click Copy, right-click Chapter4.gdb, click Paste > OK. ArcCatalog creates the copy, Cities_1.

2 Right-click Cities_1, and click Delete > Yes.

YOUR TURN

Open a Computer window, browse to Chapter4.gdb in the Chapter4 folder of MyExercises, right-click Chapter4.gdb to get its properties and size, and take a look at the files inside of it comprising the cities, tracts, and census tract data. You should find that the folder size is approximately 1.5 MB on the disk and that its 55 files are incomprehensible. You need ArcCatalog or Catalog in ArcMap to use, manipulate, and understand these files. Leave the Computer window open for use in the following steps:

Compress a file geodatabase

1 In the left panel of ArcCatalog, right-click Chapter4.gdb, click Administration > Compress File Geodatabase > OK.

2 Use a Computer window to check the size of the Chapter4.gdb folder. Verify that the compressed file geodatabase is a little less than 1 MB on the disk, about a 40 percent reduction in size.

While ArcMap can process compressed feature layers by uncompressing them on the fly, you will use the next step to uncompress the folder and get the layers back to original size.

3 In the left panel, right-click Chapter4.gdb, click Administration > Uncompress File Geodatabase > OK.

4 Close ArcCatalog.

YOUR TURN

Open ArcMap and create a new map document called **Tutorial4-3.mxd** with relative paths stored in the Chapter4 folder of MyExercises. Add the two layers and table from Chapter4.gdb and symbolize as seen in the following. Save your map document.

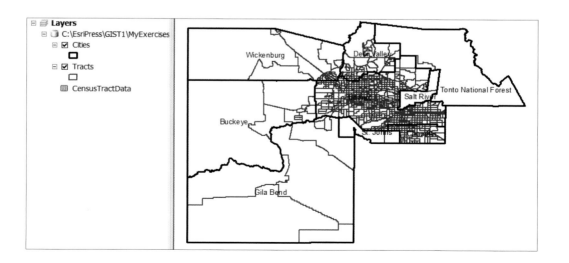

Tutorial 4-3

Modifying an attribute table

Most of what gets processed or displayed in a GIS depends on attribute table values. There is much, therefore, that you need to know about working on tables, including deleting, modifying, and creating columns.

Delete unneeded columns

Many map layers have extra or unnecessary attributes, from the user's point of view, that you can delete to tidy up.

1 In the table of contents, right-click Tracts and click Open Attribute Table. It's not possible to delete the primary key created by ArcGIS, ObjectID, or the Shape attribute. GEOID10 is the tract geocode that you need for joining the CensusTractData table to the tracts map layer, but all other fields are candidates for deletion.

2 In the table, right-click the header for the STATEFP10 column, click Delete Field, and click Yes.

3 Similarly delete the following fields: COUNTYFP10, TRACTCE10, NAME10, NAMELSAD10, MTFCC10, and FUNCSTAT10.

4 Close the Tracts table.

> ### YOUR TURN
> GEOID10 is a unique city identifier in the United States. Delete fields from the Cities layer to keep just OBJECTID, Shape, GEOID10, NAME10, Shape_Length, and Shape_Area. When finished, close the Cities table.

Modify a geocode

It is often necessary to join two tables to make a single combined table. For example, there are thousands of census variables, so it is impractical to have all needed census variables for tracts stored in the Tracts attribute table. Instead, you select the variables you wish, download a corresponding table (which includes tract geocodes) from the Census Bureau

4-1
4-2
4-3
4-4
4-5
4-6
A4-1
A4-2

website, and join the table to the tract polygon table by geocode. The GEOID10 column of the Tracts attribute table and the GEOid2 column of the CensusTractData table are the corresponding geocodes for these tables. These attributes would match, except that GEOID10 has a text data type while GEOid2 has a numeric data type. You can tell because text data is left-aligned in its column while numeric data is right-aligned.

1 In the table of contents, right-click CensusTractData, click Open, and sort GEOid2 ascending. You can see that this numeric attribute's first sorted value is 4013010101.

2 Close CensusTractData. Similarly, open the Tracts attribute table and sort GEOID10 ascending. The first sorted value is the text value 04013010101, with a leading zero, but otherwise matches the geocode in CensusTractData.

To get this table's geocode to join with that in CensusTractData, all you have to do is create a numeric version of it, which is easy with the Field Calculator. The numeric version drops the leading zero.

3 In the Tracts table, click the Table Options arrow and click Add Field.

4 In the Fields dialog box, type **GEOID10Num** in the Name field, change the Type to Double, and click OK. You need Double, which can have values up to 15 digits in length, to store GEOID10.

5 Right-click the GEOID10Num header and click Field Calculator > Yes.

6 In the Field panel, double-click GEOID10 (to set GEOID10Num = GEOID10), and click OK. That provides the needed geocode for joining the CensusTractData table to the Tracts attribute table.

7 Close the Tracts table and open the CensusTractData table.

Before making the join, first you create some new calculated columns in the CensusTractData table needed for a map document.

Calculate a new column

The CensusData table has population and per capita income for the total population, whites, Native Americans, and Hispanics. Desired for mapping are two ratios of per capita incomes for Native Americans divided by whites and for Hispanics divided by whites.

1 In the CensusTractData table, click the Table Options arrow and click Add Field.

2 In the Add Field window, type **RNatWht** in the Name field, change the Type to Float, and click OK.

The new column will contain the ratio of Native American per capita income to white per capita income. Wherever this ratio is greater than one, Native Americans on average earn more than whites. First, however, you must select only records where PCIncWht is greater than zero, because PCIncWht is the divisor for this ratio and is used to calculate values for RNatWht. Anything divided by zero is undefined, so this case must be avoided; this is done through the selection.

3 In the CensusTractData table, click the Table Options arrow and click Select By Attributes.

4 In the Select By Attributes dialog box, scroll down the list of fields, double-click WhtPCIncome to add it to the lower Select panel, click the > Symbol button, click Get Unique Values, and double-click 0 in the Unique Values list. Those actions create the expression "WhtPCIncome" > 0.

5 Click Apply > Close.

6 Right-click the RNatWht header and click Field Calculator > Yes.

7 In the Field Calculator window, double-click NatPCIncome in the Fields panel, click the / button, double-click WhtPCIncome in the Fields panel, and click OK. Rows with values for both the numerator and denominator get data values, those with a denominator that is positive but no numerator get a zero, and those with 0 or null for the denominator retain the null value.

YOUR TURN

Repeat the previous steps to calculate a new column in the CensusTractData table called **RHisWht**, which is the ratio of HispPCIncome divided by WhtPCIncome. This is the ratio of per capita income of Hispanics divided by the per capita income of whites. Clear the selection and close the table when finished. Save your map document.

4-1
4-2
4-3
4-4
4-5
4-6
A4-1
A4-2

Tutorial 4-4

Joining tables

*Next, you will join the CensusTractData table to the polygon Tracts feature class.
The same steps work if your map layer is a shapefile or map layer in another format
supported by ArcMap.*

1 Save your map document as **Tutorial4-4.mxd** to the Chapter4 folder.

2 In the ArcMap table of contents, right-click
the Tracts layer, click Joins and Relates >
Join.

3 Make the selections shown in the
image on the right.

4 Click OK.

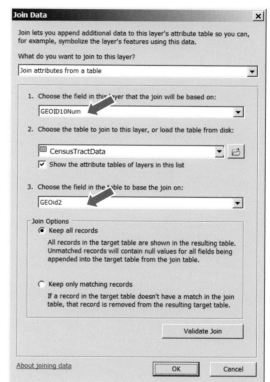

5 Open the Tracts attribute table, scroll to the right in the table, and verify that ArcMap
joined the CensusTractData table to the Attributes of Tracts table.

	Shape_Length	Shape_Are	GEOID10Num	OBJECTID *	GEOid2 *	PopTotal	PopWhite	PopNative
	0.127403	0.000755	4013422644	712	4013422644	7287	6319	33
	0.09817	0.000499	4013422643	711	4013422643	5789	4929	41
	0.063555	0.000249	4013422642	710	4013422642	5715	4888	41
	0.132432	0.000748	4013422641	709	4013422641	6346	5052	54
	0.170811	0.001141	4013815900	889	4013815900	5981	4533	23
	0.075694	0.000274	4013815800	888	4013815800	4083	3279	45

Tracts

6 Close the Tracts table.

YOUR TURN

Now that you have census tract data attached to the census tract map layer, you can make a quick map of an interesting variable. Are there places in Maricopa County where Native Americans typically have higher per capita income than whites? Make a choropleth map using the RNatWht (ratio of Native American per capita income divided by white per capita income) attribute you created. Use quantiles for the numeric scale and the red to yellow to green color ramp. When finished, save your map document.

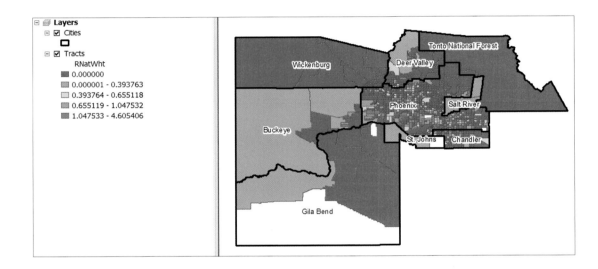

Tutorial 4-5

Creating centroid coordinates in a table

The centroid of a polygon is the point at which the polygon would balance on a pencil point if it were cut out of cardboard. A map layer consisting of polygon centroid points has many uses; for example, together, polygons and their centroids give you the ability to display two attributes of the same map layer in the same map, one as a choropleth map and the other as a size-graduated point marker map. Next, you use ArcMap's tool for calculating polygon centroids.

Add x,y coordinates to a polygon attribute table

As a precaution and as a simplification, when you add new fields to an attribute table that has a joined table, it's a good idea to remove the join first, create the new fields, and then rejoin the table.

1 Save your map document as **Tutorial4-5.mxd** to the Chapter4 folder.

2 Right-click Tracts in the table of contents, click Joins and Relates > Remove Joins > Remove All Joins. Your choropleth map disappears because the attribute that was displayed was in the joined table, which you just removed.

3 Open the Tracts table, click the Table Options arrow, and click Add Field.

4 Type **X** as the Name, select Double for Type, and click OK.

5 Repeat steps 3 and 4 except call the new field **Y**.

6 Right-click the X header, click Calculate Geometry > Yes. Examine the Calculate Geometry window and click OK > Yes.

7 Repeat step 6 except right-click the Y header and select Y Coordinate of Centroid. Close the attribute table when finished.

Shape_Length	Shape_Area	GEOID10Num	X	Y
0.127403	0.000755	4013422644	-111.598551	33.366828
0.09817	0.000499	4013422643	-111.618524	33.357267
0.063555	0.000249	4013422642	-111.627432	33.371808
0.132432	0.000748	4013422641	-111.661659	33.357603
0.170811	0.001141	4013815900	-111.757895	33.264
0.075694	0.000274	4013815800	-111.697149	33.284485

YOUR TURN

Rejoin the CensusTractData table to the Tracts layer.

Export a table

When you export joined tables as a table, you get all the attributes of the joined tables stored as one table permanently. (Similarly, if you export a map layer that has a joined table, the new map layer's attribute table has both tables joined permanently.) There are several possible uses for the new table, one of which is to use it to make a new point layer based on the centroid coordinates.

1 Open the Tracts attribute table, click the Table Options arrow > Export.

2 In the Output table field of the Export Data window, click the browse button, change the Save As type to File Geodatabase tables, browse to the Chapter4 folder, double-click Chapter4.gdb, change Name to **TractCentroids**, and click Save > OK > Yes. Open the table to see that it has all of the columns of both joined tables. Then close the table.

3 Close the Tracts attribute table.

Create a feature class from an XY table

1 On the Menu bar, click Windows > Catalog. This opens a version of ArcCatalog as a window in ArcMap, thus providing quick access to GIS utility programs.

2 Expand the Chapter4.gdb file geodatabase.

3 Right-click TractCentroids, click Create Feature Class > From XY Table.

4 In the Create Feature Class From XY Table window, click the Coordinate System of Input Coordinates button, click the Add Coordinate System button 🌐 ▼ > Import, browse to Chapter4.gdb, double-click Tracts in Chapter4.gdb, and click OK.

The coordinate system's geographic spherical coordinates are the same as those of the Tracts layer, so the simplest option is to import the system specification from Tracts.

5 Click the browse button for Output, change the Save As type to File and Personal Geodatabase feature classes, browse to and double-click Chapter4.gdb, change the Name to **CensusTractCentroids**, click Save > OK.

6 Drag CensusTractCentroids from the Catalog window to the map document and drop it at the top of the table of contents.

7 Close the Catalog window in ArcMap.

8 Open the Cities attribute table, click the row selector for Phoenix to select that record and polygon on the map, and close the table.

9 Right-click Cities in the table of contents, click Selection > Zoom to Selected Features. Now you can get a better look at the centroids point layer you just created.

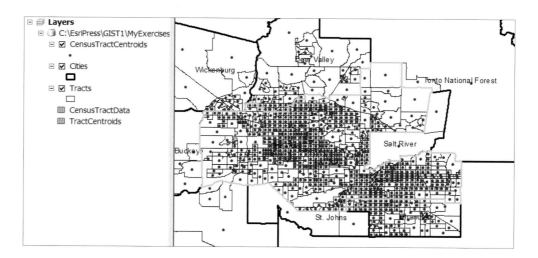

Symbolize a choropleth and centroid map

Here you symbolize a map using both the tract polygons and centroids. Let's see how per capita income compares with percentage of total population that is Hispanic.

1 In the table of contents, right-click the Tracts layer, click Properties > the Symbology tab.

2 In the Show panel, click Quantities > Graduated Colors. Under Fields, select PopHispanic for the Value field, change Normalization to PopTotal, click Classify, change Method to Quantile, and click OK. Change the color ramp to yellow through brown.

3 In the Symbology tab, click the Label header to the right of the Symbol and Range headings, click Format Labels, click the Numeric Category, change the number of decimal places to 2, and click OK twice.

4 Right-click CensusTractCentroids in the table of contents, click Properties > the Symbology tab.

5 In the Show panel, click Quantities > Graduated Symbols. Under Fields, change Value to PCIncome.

6 Change the Symbolize Size range to 2 to 10.

7 Click the Template button, choose Circle 2, and click OK.

8 In the Classification panel, change the number of classes to 5, click Classify, change Method to Quantile, and click OK.

9 In the Symbology tab, click the Label header, click Format Labels, click the Numeric Category, change the number of decimal places to 0, click OK twice, and save your map document. Now you can plainly see that Phoenix, Arizona has areas with concentrations of Hispanic population, and those areas tend to be low income.

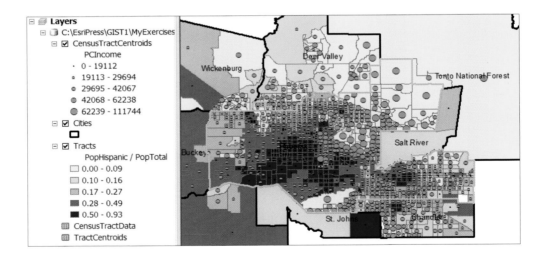

Tutorial 4-6

Aggregating data

A car beat is the patrol area of a single police car. This tutorial has you count (or aggregate) eating and drinking businesses in Rochester, New York within car beats. Such businesses are crime attractors, so it is useful to have a map showing where they are concentrated. In this case, the workflow has four steps. First, you join a code table to all business points in order to identify the businesses of interest. Second, you apply a definition query to businesses, limiting them to eating and drinking businesses. Third, you use a spatial join of car beats to eating and drinking businesses to count them up by car beat. Finally, you use the results to create a choropleth map of car beats representing the number of drinking and eating places in each car beat.

Join a code table to an attribute table

Here you do a one-to-many join. Earlier in this chapter, you did a one-to-one join between the CensusTractData table and the Tracts attribute table. Each CensusTractData record had one and only one matching Tract polygon and record. This time, each SIC (standard industrial code) record in the code table potentially has many matching records in the Businesses point layer; for example, all of the drinking places in Rochester for the SIC record for code 5813.

1 Open Tutorial4-6.mxd from the Maps folder. The map that opens displays police car beats in Rochester as polygons, and all businesses as points.

2 Save your map document to the Chapter4 folder.

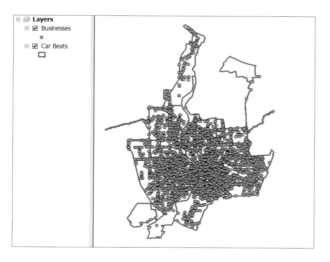

3 Open the Businesses attribute table and examine its attributes. There are 9,325 businesses of all kinds with the name and address of each business along with a four-digit SIC code. To break this code you need the code table called SIC. Notice that the SIC column in the Businesses attribute table has the text data type.

OBJECTID *	Shape *	NAME	ADDRESS	SIC
9008	Point	WATER-WISE INC	311 EXCHANGE BLVD	7389
9006	Point	WATER STREET GRILL	179 N WATER ST	5812
9007	Point	WATER STREET MUSIC HALL	204 N WATER ST	5813
9009	Point	WATERPRO SUPPLIES CORP	609 BUFFALO RD	9999
9010	Point	WATERWORKS J WELLINGTONS	315 ALEXANDER ST	5812
9011	Point	WATKINS ADVERTISING	105 LANARK CRES	9999

4 Close the Businesses attribute table. Click the List By Source button at the top of the table of contents and open the SIC table. This code table has two-, three-, and four-digit SIC codes and descriptions. The two-digit code, 01 for Agriculture, is broken down into three-digit categories, for example 011 Cash grains. Then three-digit codes are broken down into four-digit business types, for example, 0111 Wheat. Note that SICCODE also has the text data type, which makes the join to Businesses possible.

OBJECTID *	SICCODE *	SICDESCR
1084	5736	Musical instrument stores
1085	58	Eating And Drinking Places
1086	581	Eating and Drinking Places
1087	5812	Eating places
1088	5813	Drinking places
1089	59	Miscellaneous Retail

5 Close the SIC table. Right-click the Businesses layer in the table of contents, click Joins and Relates > Join and type or make the selections shown in the graphic on the next page.

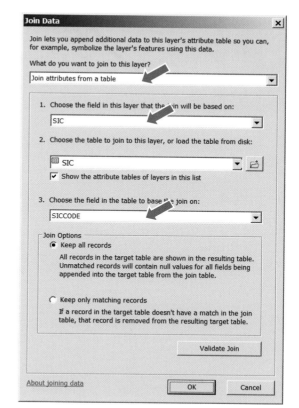

6   Click the Validate Join button. You get a report that 9,253 of 9,325 business records successfully join. If you needed to diagnose and repair the non-matches, you could, but you skip that here.

7   In the Join Validation window click Close and click OK.

8   Open the Businesses attribute table, scroll to the right, and see the joined SIC code descriptions. Of course, all irrelevant SICCODE values, including those two- and three-digit values, were ignored in the join process. Only relevant four-digit codes were used.

9   Close the attribute table.

Create a definition query

1   Open the Businesses layer's properties and click the Definition Query tab.

2   Click the Query Builder button, and build the query criterion: **SIC.SICDESCR = 'Eating places' OR SIC.SICDESCR = 'Drinking places'** using the Get Unique Values button to get description values.

3   Click OK > OK. There are 457 such places now displayed.

Use a spatial join to count eating and drinking places by car beat

If you do a spatial join of a polygon layer with a point layer, the result is statistics by polygon for the points. Each such join automatically includes a count of points per polygon, and if there are quantitative attributes of the points they can have statistics included, as seen in the graphic for step 1.

1 Right-click Car Beats, click Joins and Relates > Join, and make selections as follows:

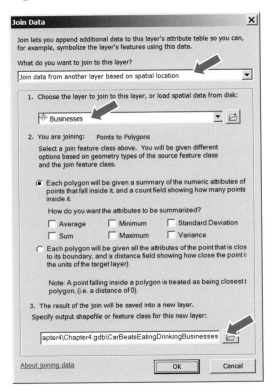

2 Click OK.

3 Remove the original car beats layer from the table of contents.

4 Symbolize the choropleth map using the Count_ attribute of CarBeatsEatingDrinkingBusinesses. Use five quantiles and label each polygon with the Beat attribute. The result is a good, high-level map for scanning Rochester for car beats with many eating and drinking places. A good interactive version of this map would have the new layer turn off when zoomed in and the Businesses point layer turned on for details.

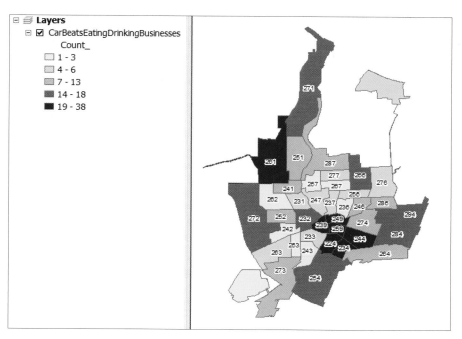

5   Save your map document and close ArcMap.

Assignment 4-1

Investigate educational attainment

In this assignment, you carry out all of the standard steps to work with census data in a GIS. Chapter 5 covers downloading and preparing data from the Census website, so this assignment starts at the point after census data is downloaded and ready for use. You create a file geodatabase, import a census tract map and data table into the file geodatabase, carry out several data utility and calculation steps, join the data table to the tract map, and produce a map layout. The data is on education attainment in Allegheny County, Pennsylvania. Your objective is to create a layout comparing the fraction of males versus females 25 or older who have an associate college degree or higher. You can imagine that spatial patterns are similar, but is high educational attainment the same for both sexes?

Get set up

First, rename your assignment folder and create a map document.

- Rename the folder \EsriPress\GIST1\MyAssignments\Chapter4\Assignment4-1YourName\ to your name or student ID. Store all files that you produce for this assignment in this folder.

- Create a new map document called **Assignment4-1YourName.mxd** with relative paths.

Use the following steps to project the table of contents layer to the state plane coordinate system, with coordinates in feet:

- Right-click Layers in the table of contents, click Properties > Coordinate System tab.

- Expand Projected Coordinate Systems, State Plane > NAD 1983 (US feet) and click NAD 1983 State Plane Pennsylvania South FIPS 3702 (US Feet) > OK.

Build the map

Create a new file geodatabase called **Assignment4-1YourName.gdb**. Import data into it:

- Import \EsriPress\GIST1\Data\DataFiles\tl_2010_42003_tract10.shp as a feature class and rename it Tracts. GEOID10 is the tract geocode in this table and it has the text data type.

- Import \EsriPress\GIST1\Data\DataFiles\AllCoEdAttain.xlsx, HighEdAttainment$ worksheet as a table and rename it **EducationalAttainment**. This is 2010 census tract data for Allegheny County on educational attainment. Attributes include: GEOid2 = tract geocode, numeric data type; Female25_ = total female population 25 and older; FemaleAssociateDegree = female population 25 or older with an associate degree as highest educational attainment; FemaleBachelorDegree = female population 25 or older with a bachelor's degree as highest educational attainment; and so forth up through FemaleDoctorateDegree. Include FemaleProfessionalDegree in your calculations. Males have similar attributes.

4-1
4-2
4-3
4-4
4-5
4-6
A4-1
A4-2

Add Tracts and EducationalAttainment to your map document.

Prepare Tracts for use:

- Create a new geocode, matching the data type of the geocode in the data table.
- Clean up the attribute table, deleting unneeded attributes.

Create two new attributes in EducationalAttainment:

- **PCollegeF** = Percentage of females 25 and older who have an associate college degree or higher (so values possibly could range between 0 and 100). Use Float as the data type.
- **PCollegeM** = Percentage of males 25 and older who have an associate college degree or higher. Use Float as the data type.
- *Hint:* Before calculating PCollegeF, select rows where Female25_ > 0. Otherwise you will get a warning message that the calculation failed (in cases where the denominator of your expression was zero). Do the same for males.

Create a layout that compares the two new attributes, PCollegeF and PCollegeM.

- Use the map layer and table in your file geodatabase (and not the shapefile or Excel table).
- Use two data frames and two separate choropleth maps, one for each new attribute.
- Use the same color ramp and numeric scale for both maps and include a single, common legend.

Hint: Finish one data frame. Then copy and paste it for the second data frame, making modifications as necessary. Create a layer file for the first data frame's map and use it to symbolize the second data frame's map.

Create a scatterplot of PCollegeF versus PCollegeM and add it to the layout.

- Use View > Graphs > Create Graph to create the scatterplot. Experiment with modifying and improving your graph's axes (e.g., use tic marks of 20), add vertical grid lines, increase the sizes of fonts, etc. by right-clicking your graph and clicking Advanced Properties.
- Once the plot is created, resize its window so that the vertical and horizontal axes are the same length.
- Right-click the graph and click Copy As Graphic. On your layout, click Edit > Paste. Reposition and align the graph and other elements of your layout.
- On the layout window, click Customize > Toolbars > Draw and draw a red line from (0,0) to (100,100) on your graph. This is the line of perfect equality of educational attainment for males and females. Think about the pattern you see on your maps and the graph.
- Save your layout and export it as Assignment4-1YourName.jpg.

Assignment 4-2

Compare serious crime with poverty in Pittsburgh

The criminology literature finds that much crime is related to poverty: the larger the population living in poverty, the higher the rate of certain kinds of crimes. Let's see if this relationship is evident in Pittsburgh at the census tract level. In the process, you'll build a file geodatabase, import map layers into it, build a code table from scratch and join it to a map layer, join census data to a map layer, and carry out a spatial join to aggregate data.

Get set up

First, rename your assignment folder and create a map document.

- Rename the folder \EsriPress\GIST1\MyAssignments\Chapter4\Assignment4-2YourName\ to your name or student ID. Store all files that you produce for this assignment in this folder.
- Create a new map document called **Assignment4-2YourName.mxd** with relative paths.

Use the following steps to project the table of contents layer to the state plane coordinate system, with coordinates in feet:

- Right-click Layers in the table of contents, click Properties > Coordinate System tab.
- Expand Projected Coordinate Systems > State Plane > NAD 1983 (US feet) and click NAD 1983 State Plane Pennsylvania South FIPS 3702 (US Feet) > OK.

Build the map

Create a new file geodatabase called **Assignment4-2YourName.gdb**. Import the following data into it:

- \EsriPress\GIST1\Data\Pittsburgh\Shapefiles\PittsburghSeriousCrimes2008.shp—point shapefile of serious crime offense locations in Pittsburgh during summer, 2008. Attributes include: CCN = police ID for offense, Address = location of the offense, DateOccur = date of the offense, Hierarchy = FBI hierarchy code for the offense (1 = Criminal Homicide, 2 = Forcible Rape, 3 = Robbery, 4 = Aggravated Assault, 5 = Burglary, 6 = Larceny-theft, 7 = Motor Vehicle Theft, 8 = Arson).
- \EsriPress\GIST1\Data\Pittsburgh\City.gdb\PghTracts—polygon features of Pittsburgh 2010 census tracts.
- \EsriPress\GIST1\Data\Pittsburgh\Shapefiles\PovertyTracts.xlsx with PghPovertyTracts worksheet—2010 tract data for poverty. Attributes include: GEOid = tract geocode (numeric data type), PopWithPovStatus = population for whom poverty status is known (presumably the same as total population), PopBelowPovLevel = population who are below the poverty level.

Join PghPovertyTracts to PghTracts and make a choropleth map for population below the poverty line.

4-1
4-2
4-3
4-4
4-5
4-6
A4-1
A4-2

Create a new code table in your file geodatabase:

- Right-click your file geodatabase in Catalog and click New > Table.
- Name the table UCRHierarchy with no alias. Click Next > Next.
- Create a field called UCR with the Short Integer data type. Create another field called Crime with the Text data type and length 20.
- Click Customize > Toolbars > Editor.
- On the editor toolbar, click Editor > Start Editing > Continue.
- Open UCRHierarchy in the table of contents, type **1** for UCR and Criminal Homicide for Crime in the first row.
- Finish up inputting rows found with codes and crime types from the description of PittsburghSeriousCrimes2008.shp above. When done, click Editor in the Editor toolbar, Save Edits, Stop editing, and close the Editor toolbar. Close your new table.

Create the following query definition for PittsburghSeriousCrimes2008 and rename that layer in the table of contents to be PittsburghSeriousCrimesSummer2008:

"DateOccur" >= date '2008-06-01' AND "DateOccur" <= date '2008-08-31'.

Join UCRHierarchy to PittsburghSeriousCrimesSummer2008. Symbolize PittsburghSeriousCrimesSummer2008 with the Crime attribute using unique values. Use different shapes and colors for point markers. This layer will display only when you zoom in.

Spatially join PittsburghSeriousCrimesSummer2008 to PghTracts to get a count of serious crimes per tract in summer 2008. Call the output CrimeAggregatedByTracts. Use the results to symbolize PghTracts centroids with size-graduated point markers. *Hint:* Use Quantities, Graduated symbols for symbolizing the crime count (there is no need to create tract centroids in this case).

Set threshold scales so that when zoomed in to about a fourth of Pittsburgh or farther, PittsburghSeriousCrimesSummer2008 turns on and CrimeAggregatedByTracts turns off.

Create a layout of your own design for the map document, zoomed to full extent.

Analyze the data

Get set up to create a graph:

- Open the CrimeAggregatedByTracts property table, click the Fields tab, and give Count_ the alias Serious Crimes.
- Open the PghPovertyTracts table and give PopBelowPovLevel the alias Population Below Poverty Level.
- Join PittsburghPovertyTracts to CrimeAggregatedByTracts.

Create a scatterplot of Serious Crimes versus Population Below Poverty Level:

- Use View > Graphs > Create Graph to create the scatterplot.
- Once created, resize its window so that the vertical and horizontal axes are the same length.
- Right-click the graph and click Copy As Graphic. On your layout, click Edit > Paste. Reposition and align the graph and other elements of your layout.
- Save your layout and export it as Assignment4-2YourName.jpg.

Spatial data

Vast collections of spatial data are available from government agencies. You can readily download much of this data for free on the Internet, but before doing so it is helpful to get some background and see some of the major forms of this data. Spatial data is complex, with both vector and raster formats available in many file formats and with several attending characteristics such as coordinate system, feature or cell attribute properties, and intended map scale for application. This chapter provides a hands-on introduction to spatial data and then has you download or use samples from some of the major governmental suppliers.

Learning objectives

- *Learn about metadata*
- *Work with world projections*
- *Work with US map projections*
- *Work with systems of projections*
- *Learn about vector data formats*

- *Download US Census map layers*
- *Download US Census data*
- *Download American Community Survey data*
- *Explore web services for raster maps*
- *Use nationalatlas.gov*

Tutorial 5-1

Examining metadata

Spatial data needs much documentation for interpretation and proper use. "Metadata" is the term for such documentation. It describes the context, content, and structure of GIS data. In ArcGIS for Desktop, data providers use ArcCatalog to create and view metadata.

Open a map document

1 **Start ArcMap and open Tutorial5-1.mxd from the Maps folder.** These map layers come from two different data sources. The Pennsylvania Counties and Allegheny Tracts layers were obtained from the US Census Bureau, and the Allegheny County Municipalities layer was obtained from a local GIS organization, the Southwestern Pennsylvania Commission.

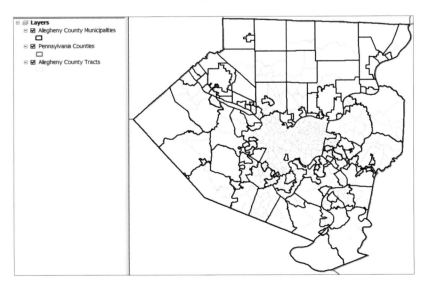

2 Save the map document to the Chapter5 folder of MyExercises.

Open a metadata file

ArcGIS metadata is described in a metadata file that can be read using ArcCatalog.

1 On the taskbar, click Start > All Programs > ArcGIS > ArcCatalog 10.1. Depending on your operating system and how ArcGIS and ArcCatalog have been installed, you may have a different navigation menu.

2 In the Catalog tree, browse to UnitedStates.gdb in the Data folder, click PACounties, and the Description tab. Be patient. It takes time for the description to appear.

3 Click Customize > ArcCatalog Options and the Metadata tab.

4 Click North American Profile of ISO19115 2003 as the Metadata style > OK.

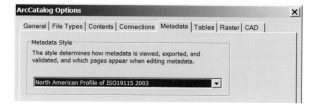

5 Scroll to the Spatial Reference option and read the entry. You learn that geographic coordinates are spherical, Geographic Coordinate System, NAD 1983 Datum.

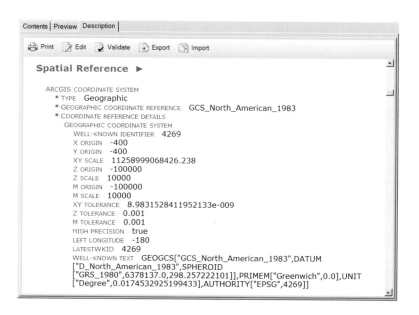

6 In the Catalog tree, browse to AlleghenyCounty.gdb in the Data folder, click Munic, click the Description tab, and wait while this loads.

7 Scroll to the Spatial Reference option and read the entry. You learn that the layer's projection is rectangular, NAD_1983_StatePlane_Pennsylvania_South_FIPS_3702_Feet. You will learn more about projections and coordinate systems in tutorials 5-2 through 5-4.

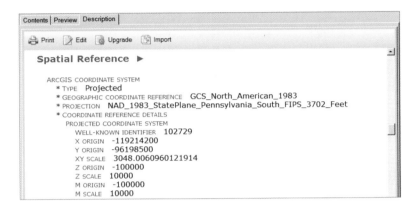

YOUR TURN

Explore some of the other metadata for Pennsylvania Counties, Allegheny Municipalities, and Allegheny County Tracts 2010, including source credits, fields, and other spatial information.

Tutorial 5-2

Working with world map projections

There are two types of coordinate systems—geographic and projected. Geographic coordinate systems use latitude and longitude coordinates for locations on the surface of a sphere while projected coordinate systems use a mathematical transformation to a flat surface and rectangular coordinates.

Set world projections

ArcMap has more than 100 projections from which you may choose. Typically, though, only relatively few are needed for most purposes.

1 Open Tutorial5-2.mxd from the Maps folder and save the map document to the Chapter5 folder.

2 Place your cursor over the westernmost point of Africa and read the coordinates on the bottom of the ArcMap window (approximately –17, 20 decimal degrees). The map and data frame are in geographic coordinates, decimal degrees, which are angles of rotation of the Earth's radius from the prime meridian at the equator. These coordinates are not intended for viewing on a flat surface such as your computer screen, so there are large distortions if you do so; for example, the north and south pole points are the horizontal lines bounding the top and bottom of the map. Instead of geographic coordinates, you should use one of the projected coordinate systems appropriate for viewing flat maps of the world. ArcMap has several and can easily project the map on the fly.

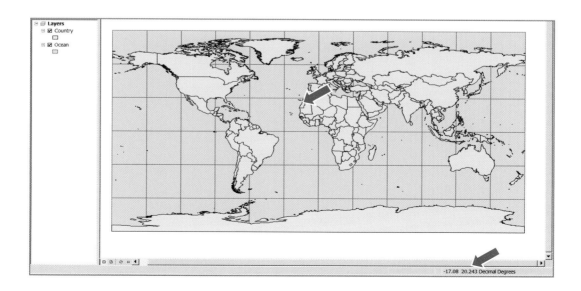

Change the map's projection to Mercator

1 In the table of contents, right-click Layers, click Properties > Coordinate System tab.

2 In the Select a coordinate system panel, scroll down to and click the plus (+) sign beside Projected Coordinate Systems, and click the plus sign beside the World folder.

3 Scroll down the coordinate systems, click Mercator (world) > OK.

4 Zoom to full extent.
The purpose of the Mercator projection is for navigation. Straight lines on the projection are compass bearings for connecting two points. This projection greatly distorts areas near the polar regions and distorts distances along all lines except the equator. The Mercator projection is a conformal projection, meaning that it preserves small shapes and angular relationships.

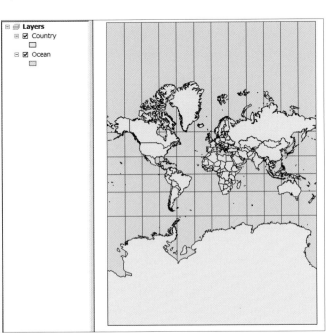

YOUR TURN

Repeat the steps of the previous exercise, but this time select the Hammer-Aitoff projection in the third step. This projection, shown in the first image below, is nearly the opposite of the Mercator. The Hammer-Aitoff is good for use on a world map, being an equal-area projection that preserves area. However, it distorts direction and distance. Repeat the steps again, this time choosing the Robinson projection. As shown in the second image below, this projection minimizes distortions of many kinds, striking a balance between conformal and equal-area projections. Save the map document.

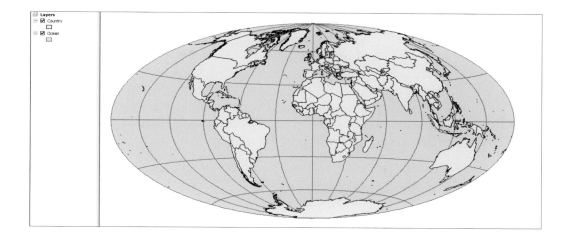

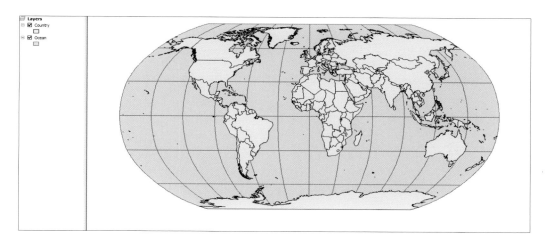

Tutorial 5-3

Working with US map projections

For most GIS applications it's good to use an equal-area projection of a large area such as the contiguous US states. Then you get accurate areas of polygons and can calculate or use accurate population densities.

Set projections of the United States

Next, you get some experience with projections commonly used for maps of the continental United States. Some projections are standard for organizations. For example, Albers Equal Area is the standard projection of both the US Geological Survey and the US Census Bureau.

1 Open Tutorial5-3.mxd from the Maps folder. Initially, the map display is in geographic coordinates. The coordinates will display as decimal degrees.

2 Save the map document to the Chapter5 folder.

3 In the table of contents, right-click Layers, click Properties > Coordinate System tab.

4 Expand Projected Coordinate Systems > Continental > North America.

5 Click North America Albers Equal Area Conic, and then click OK and Yes. Your map is now projected using the Albers Equal Area projection.

YOUR TURN

Experiment by applying a few other projections to the US map such as North America Equidistant. As long as you stay in the correct group—Continental, North America—all of the projections look similar. The conclusion is that the smaller the part of the world that you need to project, the less distortion. There remains much distortion at the scale of a continent, but much less so than for the entire world. By the time you get to a part of a state, such as Allegheny County, practically no distortion is left, as you see next. Save your map document.

Tutorial 5-4

Working with rectangular coordinate systems

For medium- and large-scale maps it's helpful to have localized projections, tuned for the study area, that have little or no discernible distortion. For this purpose, there are collections of projections. You have to check a reference map to determine which zone your study area is in and select that projection for your map.

State Plane Coordinate System

The State Plane Coordinate System is not a projection but a coordinate system dividing the 50 US states, Puerto Rico, and the US Virgin Islands into more than 124 numbered zones, each with its own finely tuned projection. Used mostly by local government agencies such as counties, municipalities, and cities, the State Plane Coordinate System is for large-scale mapping in the United States. The US Coast and Geodetic Survey developed it in the 1930s to provide a common reference system for surveyors and mapmakers. The first step in using the State Plane Coordinate System is to look up the correct zone for your area and consequently a specific projection tailored to your study area.

1 Start your web browser, go to www.ngs.noaa.gov/TOOLS/spc.shtml, and click the Find Zone link. If the link does not work, navigate to www.nsg.noaa.gov, and click Tools > State Plane Coordinates.

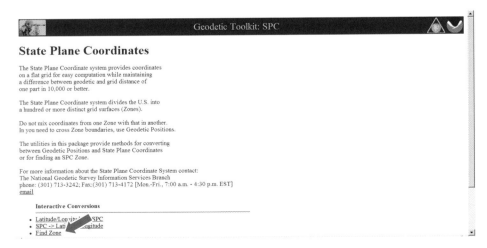

2 In the resulting web page, with the By County option button selected, click Begin.

5-1
5-2
5-3
5-4
5-5
5-6
5-7
5-8
5-9
5-10
5-11
A5-1
A5-2

3 Select Pennsylvania, click Submit, and then click Allegheny and Submit. Pennsylvania's Allegheny County is in State Plane Zone 3702 (LAMBERT | SOUTHERN ZONE).

4 Close your browser.

Add projected layers to a map document

As a default, the first map layer that you add to a map document sets the coordinate system and projection for the data frame. If all of your map layers have spatial reference data included, you should have no problem combining maps with different coordinate systems or projections. ArcMap will reproject all map layers to the data frame's projection on the fly. Next, you add a layer with a State Plane Coordinate System projection and then a layer with geographic coordinates.

1 In a new blank map, click the Add Data button, browse to the Data folder, AlleghenyCounty.gdb, and add Munic to the map. The coordinates appearing in the lower right corner of the display now appear in state plane units (feet). The origin of these coordinates (0,0) is at the lower left corner of Pennsylvania.

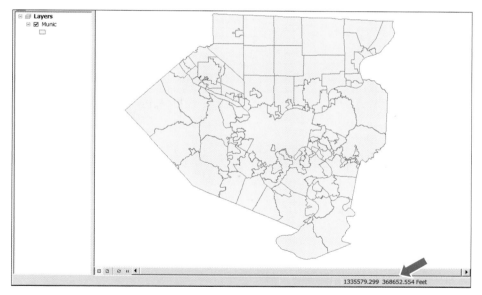

2 Save your map document as **Tutorial5-4.mxd** to the Chapter5 folder.

3 In the table of contents, right-click Layers, click Properties > Coordinate System tab. Notice that the data frame coordinates are the same as the Munic layer, NAD_1983_StatePlane_Pennsylvania_South_FIPS_3702_Feet. This is because it was the first layer added to the new map document. ArcMap will automatically project all additional layers to state plane, regardless of their original coordinates if the added layers have spatial reference data included.

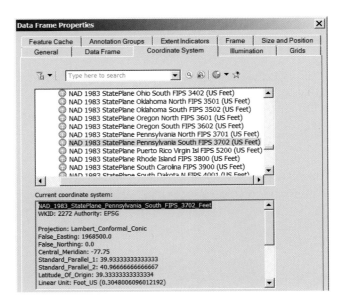

4 Click OK and change the Munic symbology to a hollow fill, black outline, width 1.15.

5 Click the Add Data button, browse to AlleghenyCounty.gdb, and add Tracts2010 to the map.

6 Change the Tracts2010 symbology to a hollow fill, light gray (20%) outline and move below Munic in the table of contents.

7 Right-click Munic in the table of contents, click Properties and the Source tab, and note the coordinate system in the Data Source panel. This layer has state plane coordinates, the same as the data frame.

8 Repeat step 7 except for Tracts2010. This layer has geographic coordinates but appears in the data frame using state plane coordinates because you added Munic first and it has state plane coordinates.

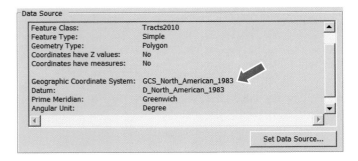

Change the data frame's coordinate system and projection to UTM

5-1
5-2
5-3
5-4
5-5
5-6
5-7
5-8
5-9
5-10
5-11
A5-1
A5-2

The US military developed the universal transverse Mercator (UTM) rectangular coordinate system in the late 1940s. It includes 60 longitudinal zones defined by meridians that are 6° wide. ArcGIS has UTM projections available for the northern and southern hemisphere of each zone. These zones, like state plane, are good for areas about the size of a state (or smaller) and have the advantage of covering the entire world.

1 Start your web browser, go to www.dmap.co.uk/utmworld.htm, **and determine the UTM zone for western Pennsylvania.** You should find that western Pennsylvania is in zone 17 north.

2 Close your browser.

3 In the table of contents, right-click the Layers data frame, and click Properties.

4 Click the General tab and note that the map units are feet.

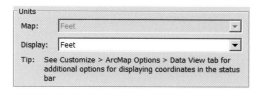

5 Click the Coordinate System tab, expand Projected Coordinate Systems > UTM > NAD 1983. Click NAD 1983 UTM Zone 17N. UTM is a metric system and thus uses meters as the map unit. Your display remains in feet so you need to set the display units to meters.

6 Click the General tab and click Meters as the display units > OK. The coordinate system and map appearance change slightly accordingly. Notice that the coordinates are now in meters.

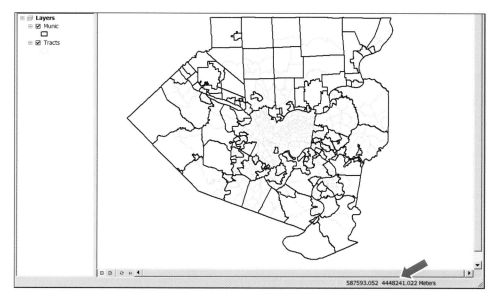

Assign a coordinate system and projection to a feature class

All GIS layers should have a coordinate system and projection defined (that is, have spatial reference data), but sometimes you will receive a shapefile or other GIS layer that does not have this data and you will need to assign this yourself. Note that the needed spatial reference data is not metadata but data that is part of the map layer. In a shapefile, it is a separate file and has a .prj extension. A common coordinate system for North American files is the Geographic Coordinate System (GCS), North American Datum 1983 (NAD 1983) projection, which is used by organizations such as the US Census Bureau and is the coordinate system of the shapefile that you are about to use.

1 In Catalog, navigate through the Data > DataFiles folder, right-click AlleghenyCountyBlockGroups.shp, click Properties > XY Coordinate System tab. The coordinate system projection for this shapefile is <Unknown>. Of course you must know the correct coordinate system to assign to the map layer from external information, metadata, or from other sources. If you assign the wrong coordinates, your map will not display correctly in your map document.

2 Expand Geographic Coordinate Systems, expand North America, click NAD 1983 and OK.

3 Repeat step 1 to see that the layer now has its coordinate system data included.

4 Add AlleghenyCountyBlockGroups to your map document. You might see a warning that the data is in a coordinate system different from that of the current data but this is OK because it has its native coordinate system assigned. Notice that the block groups overlay nicely with the other map layers that are now projected as UTM rectangular coordinates.

5 Remove AlleghenyCountyBlockGroups from the table of contents.

6 Save your map document.

Tutorial 5-5

Learning about vector data formats

This tutorial reviews several file formats commonly found for vector spatial data, other than the file geodatabase covered in chapter 4. Included are Esri shapefiles and coverages as well as computer aided design (CAD) files and XY event files.

Examine a shapefile

Many spatial data suppliers use the shapefile data format for vector map layers because it is so simple. Shapefiles appeared about the same time that personal computers became popular. A shapefile consists of at least three files: an SHP file, a DBF file, and an SHX file. Each of these files uses the shapefile's name but with the different file type extensions. The SHP file stores the geometry of the features, the DBF file stores the attribute table, and the SHX file stores an index of the spatial geometry for speeding up processing. Next, you examine AlleghenyCountyBlockGroups.shp in more detail.

1 **Examine AlleghenyCountyBlockGroups.shp in Catalog.**
It appears as an entry in one line with an icon representing a polygon map layer. ArcCatalog treats the several files as a unit and provides utilities such as renaming the shapefile in one location.

2 **Open a Computer window and navigate to the Data > DataFiles folder.** Now you can see that there are five files for the shapefile, including the projection (.prj) file that you created in tutorial 5-4 when you added a spatial reference for the layer's coordinate system. Shapefiles have three to seven associated files.

3 Close the Computer window.

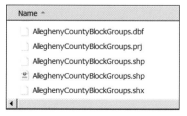

Add a coverage to ArcMap

A coverage is a legacy Esri spatial data format from times when personal computers did not even exist. Coverages typically store one or more feature classes that are related in a folder named for the coverage. For example, in a cadastral (land ownership) dataset it is common for a coverage to store the parcel boundaries as polygons and the parcel lines making up the polygons as arcs (lines). You can add coverage data to ArcMap and use it for analysis and presentation, but you cannot edit coverage data with ArcMap. When browsing data within Windows Explorer, coverages appear as folders containing several files. In the following, you can see the four coverages in the EastLiberty folder with the contents of the Building coverage appearing in the right-hand side of Windows Explorer. The Building coverage has 18 files.

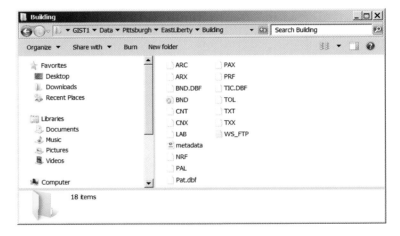

1   In a new blank map, click the Add Data button, browse through the Data > Pittsburgh > EastLiberty folder, and double-click Building.

2   Click the Polygon layer icon, click Add, and OK. A coverage behaves like any other vector layer in ArcMap. It has the same appearance and has an attribute table.

3   Save the map document as **Tutorial5-5.mxd** to the Chapter5 folder.

> ### *YOUR TURN*
>
> Add Curbs Arc and Parcel Polygon coverages to your map for the East Liberty neighborhood. Change the symbols to your liking and examine the attribute table for the Parcel layer.

Convert a coverage to a feature class

If you need to edit the attribute tables or geometry of a coverage, you must export it first as a shapefile or file geodatabase feature class.

1 In the table of contents, right-click the Building Polygon layer, click Data > Export Data.

2 Browse to Chapter5.gdb in the Chapter5 folder of MyExercises and save the output feature class as **Building**.

3 Click OK > Yes to add the exported data to the map as a layer. Now you could edit the polygons in the Building feature class and add the missing spatial reference for the layer. You learn about editing features in chapter 7.

4 Save the map document.

Add a CAD drawing file

Many organizations have CAD (computer aided design) drawings that you can display in ArcMap in their native format. ArcMap can add CAD files in one of two formats: as native AutoCAD (.dwg) or as Drawing Exchange Files (.dxf) that most CAD software can create. When viewed in Catalog, a CAD dataset appears with a light blue icon. An AutoCAD file is much like a coverage in that it has different kinds of vector features in the same file. You can see CampusMap.dwg in ArcCatalog in the following:

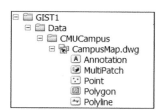

1 In a new blank map, click the Add Data button, browse through the Data > CMUCampus folder, click the CampusMap.dwg icon, click Add, and OK to the spatial warning message. The following map of the Carnegie Mellon University campus appears in ArcMap as a layer group that contains many feature types, including lines, polygons, and text. This map is from the Facilities Management Department and has spatial reference data.

5-1
5-2
5-3
5-4
5-5
5-6
5-7
5-8
5-9
5-10
5-11
A5-1
A5-2

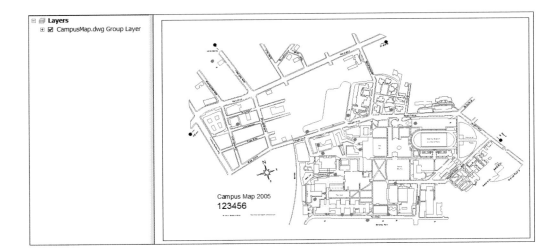

2 In the table of contents, expand the CampusMap.dwg Group Layer, and turn off all layers except CampusMap.dwg Polygon. This shows academic and residential buildings on CMU's campus.

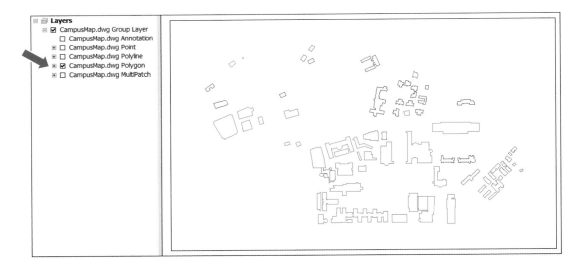

3 Right-click the CampusMap.dwg Polygon layer, click Properties > Drawing Layers tab. Notice that you can turn layers on and off. You can also symbolize the layers using the Symbology tab.

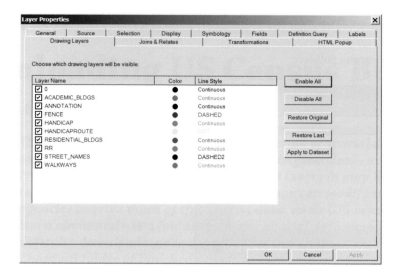

4 Close the Layer Properties window and remove CampusMap.dwg Group Layer from the table of contents.

Export GIS features to CAD

Sometimes you may need to deliver GIS features to a person working with CAD software. You can export features to AutoCAD (.dwg) or Drawing Exchange Files (.dxf) formats, which can then be opened by most commercial CAD applications.

1 Click the Add Data button, browse through the Data folder to 3DAnalyst.gdb, click Topo, and click Add. This is two dimensional topography.

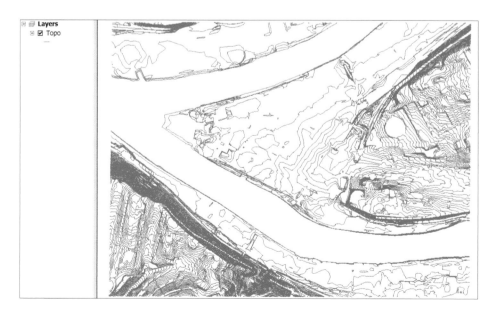

Process data in Microsoft Excel

The census data that you downloaded in the previous exercise as a text table needs some cleaning up using software such as Microsoft Excel before using in your GIS.

1 Open Microsoft Excel on your computer, click File > Open, select All Files (*.*) for Files of type, browse to the Chapter5 folder, and double-click the CSV (comma separated values) file DEC_10_SF1_QTP1.

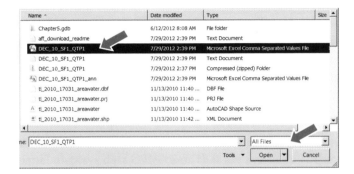

2 Delete column C and then columns F–KC.

3 Rename the columns **Id**, **Id2**, **Both sexes**, **Male**, and **Female**, as shown in the figure below. The resulting table shows only geocodes (Id and Id2) and basic population fields. One problem is that the fields for the join attributes do not match in data type and therefore cannot join. In the TIGER tracts polygon feature class, the join field (GEOID10) is a text field and in this table the corresponding join field (Id2) is a number field. You can tell because data in text fields is left-aligned and that in numeric fields is right-aligned. A simple solution is convert the number field, Id2, to text.

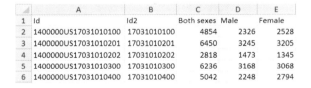

	A	B	C	D	E
1	Id	Id2	Both sexes	Male	Female
2	1400000US17031010100	17031010100	4854	2326	2528
3	1400000US17031010201	17031010201	6450	3245	3205
4	1400000US17031010202	17031010202	2818	1473	1345
5	1400000US17031010300	17031010300	6236	3168	3068
6	1400000US17031010400	17031010400	5042	2248	2794

4 Click the B column selector cell to select that column. On the Home ribbon in Excel, click the Find & Select button and click Replace. Type values as shown (note the single quotation mark in the "Replace with" text box).

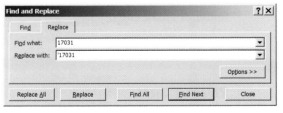

5 Click Replace All. All cells get a note, indicated by a small green triangle, that numbers are stored as text.

	A	B	C	D	E
1	Id	Id2	Both sexes	Male	Female
2	1400000US17031010100	17031010100	4854	2326	2528
3	1400000US17031010201	17031010201	6450	3245	3205
4	1400000US17031010202	17031010202	2818	1473	1345
5	1400000US17031010300	17031010300	6236	3168	3068
6	1400000US17031010400	17031010400	5042	2248	2794

6 At the bottom of the spreadsheet, double-click the lower left tab that has the text DEC_10_SF2_QTP1, and type **TractsPop2010** to rename the tab.

	A	B	C	D	E
1	Id	Id2	Both sexes	Male	Female
2	1400000US17031010100	17031010100	4854	2326	2528
3	1400000US17031010201	17031010201	6450	3245	3205
4	1400000US17031010202	17031010202	2818	1473	1345
5	1400000US17031010300	17031010300	6236	3168	3068
6	1400000US17031010	17031010400	5042	2248	2794

TractsPop2010

7 Save your file as an Excel Workbook called **CookCountyTractPop2010.xlsx** to the Chapter5 folder.

8 Close Excel.

Import files into a file geodatabase

Before adding your features and tables to ArcMap, you will import them into Chapter5.gdb. Excel spreadsheet attribute tables cannot be edited directly in ArcMap so they must be added via a file geodatabase.

1 Launch ArcMap with a new empty map document.

2 Save your map document as **Tutorial5-7** to the Chapter5 folder.

3 In Catalog, right-click Chapter5.gdb, click Import > Feature Class (multiple...) and make the selections as shown in the graphic to the right.

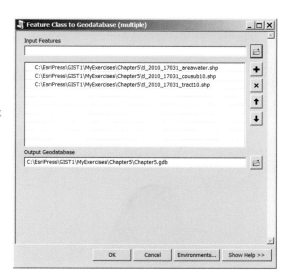

4 Click OK and wait for the shapefiles to import into the file geodatabase. In Catalog, right-click Chapter5.gdb, click Import > Table (single…) and make the selections as shown in the following: TractsPop2010$ is the worksheet within the Excel spreadsheet with the population data.

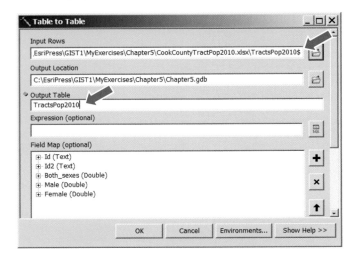

5 Click OK and wait for the worksheet from Excel to import into the file geodatabase. The table is automatically added to the map document.

Add and symbolize features

1 Add polygon features tl_2010_17031_tract10, tl_2010_17031_cousub10, and tl_2010_17031_areawater to your map document from Chapter5.gdb.

2 In the table of contents, rename the data frame **Cook County Illinois**.

3 Rename the additional layers **Tracts 2010**, **County Subdivisions**, and **Lakes and Rivers**.

4 Symbolize County Subdivisions as a thick, hollow outline and move that layer to the top of the table of contents.

5 Symbolize Lakes and Rivers as a light blue color with no outline.

Add and Join tables

Next you join the census data in TractsPop2010 to the Tracts 2010 attribute table.

1 In the table of contents, right-click Tracts 2010, click Joins and Relates > Join, and type or make selections as follows:

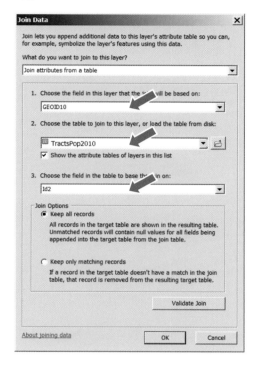

2 Click OK > Yes.

3 Open the Cook County Tracts attribute table and verify that the census data attributes (Both_sexes, Male, and Female) are joined.

	AWATER10	INTPTLAT10	INTPTLON10	Shape_Length	Shape_Area	OBJECTID *	Id	Id2 *	Both_sexes	Male	Female
▶	0	+41.8320943	-087.6818822	0.050004	0.00009	1282	1400000US17031840300	17031840300	3950	1995	1955
	0	+41.8445748	-087.6491915	0.047528	0.00009	1281	1400000US17031840200	17031840200	2338	1179	1159
	0	+41.8510058	-087.6350978	0.048332	0.000124	1287	1400000US17031841100	17031841100	7254	3458	3796
	0	+41.8555618	-087.6833420	0.033637	0.000068	1288	1400000US17031841200	17031841200	5262	2692	2570
	0	+41.8704157	-087.6750794	0.049078	0.000126	1266	1400000US17031838200	17031838200	1578	748	830
	15346	+41.9976033	-088.0176885	0.077918	0.000307	684	1400000US17031770201	17031770201	5894	2757	3137
	23682	+42.0248258	-088.0515840	0.054272	0.000181	804	1400000US17031804610	17031804610	2320	1006	1314

Tracts 2010

1 ▸ (0 out of 1319 Selected)

Tracts 2010

4 Close the table.

YOUR TURN

Using the newly joined fields, create a choropleth map showing the percentage of females by census tract. Copy the Tracts2010 layer with the joins to create a layer showing the percentages of males in Cook County, Illinois. Save your map document.

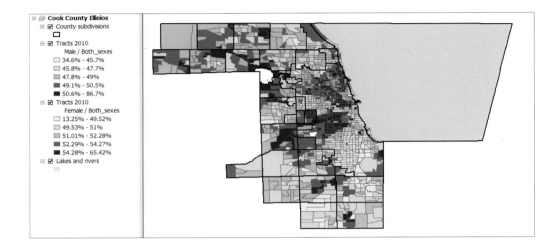

Tutorial 5-8

Downloading and processing American Community Survey (ACS) Census data

5-1
5-2
5-3
5-4
5-5
5-6
5-7
5-8
5-9
5-10
5-11
A5-1
A5-2

For more detailed data on education, income, transportation, and other subjects, you download estimates from the ACS tables. In the 2000 and previous censuses such data was collected in the census long-form along with the SF 1 data, randomly from one out of six households, and was called SF3 (Summary File 3) data. Now the Census Bureau collects ACS data monthly—approximately 3 million housing units receive a survey similar to the old long-form questionnaire. Annual, three-year, and five-year estimates produced from ACS samples are available. To learn more about the ACS, visit the website at www.census.gov/acs/www/.

In this tutorial you download ACS data at the tract level for a county. The approach you take is to first select the topic (educational attainment) then the geography and location (tracts in Cook County, Illinois). American Community Survey data can also be downloaded by block groups but is currently available only at the following website: http://www.census.gov/acs/www/data_documentation/summary_file/ *or from the US census DataFerret,* http://dataferrett.census.gov/.

Download ACS data by tract

1 Save your map document as **Tutorial5-8.mxd** to the Chapter5 folder.

2 In your web browser, go to http://factfinder2.census.gov, click the Advanced Search tab, and clear selections that you might have from the previous tutorial.

3 Click the Topics link on the left, and then within the Topics panel expand People, expand Education, and click Educational Attainment. Note that the number in parentheses might be different from the one in the figure.

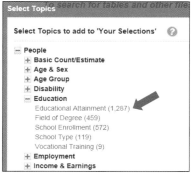

4 Close the Select Topics window.

5 Click the Geographies link on the left and the Name tab.

6 In the Enter a geography name box type **Cook County tracts**, click Go, select All Census Tracts within Cook County, Illinois, and click Add. That adds this geography to the Your Selections panel.

7 Close the Select Geographies window.

8 In the Search results large panel on the right, navigate to and click S1501 EDUCATIONAL ATTAINMENT 2010 ACS 5-year estimates > Download > OK > Download> Open > Extract all files, and extract to the Chapter5 folder. The downloaded files will be text files ACS_10_5YR_S1501 and aff_download_readme_ann, and comma separated files ACS_10_5YR_S1501_metadata.csv and ACS_10_5YR_S1501_with_ann.csv. Note that these are 2010 estimates, the data available at the time this tutorial was written. Newer estimates might be available but this tutorial uses 2010 data.

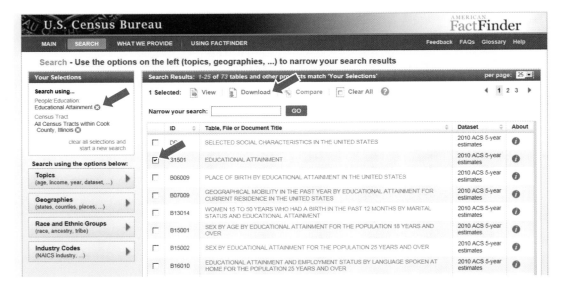

9 Close your browser.

Process data in Microsoft Excel

The census data that you downloaded in the previous steps needs cleaning up using software such as Microsoft Excel before using in GIS.

1 Open Microsoft Excel on your computer, click Open, select All Files (*.*) for Files of type, browse to the Chapter5 folder, and double-click ACS_10_5YR_S1501_with_ann.csv.

Next you need to mark the columns that you will keep so that you can delete other columns. Definitions for these columns are found in the metadata file called ACS_10_5YR_S1501_metadata.

2 Click the column heading cell A to select that column, click the color fill drop-down list under the Home tab, select a color (yellow).

3 Similarly, click the column heading buttons and use color fill for columns B, D, and J.

	A	B	C	D	E	F	G	H	I	J	K
1	GEO.id	GEO.id2	GEO.display-label	HC01_EST_VC01	HC01_MC	HC02_EST_VC01	HC02_MOE_VC01	HC03_EST_VC01	HC03_MO	HC01_EST_VC02	HC01_MO HC02
2	1400000US17031010100	17031010100	Census Tract 101, Cook County, Illinois	549	237	95	78	454	206	33	19.4
3	1400000US17031010201	17031010201	Census Tract 102.01, Cook County, Illinois	614	237	229	122	385	213	7.3	8.6
4	1400000US17031010202	17031010202	Census Tract 102.02, Cook County, Illinois	164	136	61	67	103	78	41.5	18.8
5	1400000US17031010300	17031010300	Census Tract 103, Cook County, Illinois	384	175	230	118	154	108	12.5	13.5
6	1400000US17031010400	17031010400	Census Tract 104, Cook County, Illinois	1439	468	600	256	839	304	4.1	4.5
7	1400000US17031010501	17031010501	Census Tract 105.01, Cook County, Illinois	477	226	247	146	230	134	28.5	19.1
8	1400000US17031010502	17031010502	Census Tract 105.02, Cook County, Illinois	938	466	465	406	473	250	10.9	10.1
9	1400000US17031010503	17031010503	Census Tract 105.03, Cook County, Illinois	573	367	142	133	431	270	0	5.5
10	1400000US17031010600	17031010600	Census Tract 106, Cook County, Illinois	749	231	344	198	405	168	4.8	5.8
11	1400000US17031010701	17031010701	Census Tract 107.01, Cook County, Illinois	288	123	161	113	127	72	23.6	24
12	1400000US17031010702	17031010702	Census Tract 107.02, Cook County, Illinois	452	172	219	112	233	95	9.7	11.2

ACS_10_5YR_S1501_with_ann

4 Delete all columns except the ones you marked as yellow.

5 Select columns A–D, click the color fill drop-down list > No Fill.

6 In row 1, type the column names shown in the following figure.

7 Select column B and on the Microsoft Excel Home ribbon, click the Find & Select button, click Replace, and replace values 17031 with '17031. The single quote before 17031 is essential. This converts column B to text values.

8 Type **ACSLessThanHighSch2005-10** as the sheet name.

	A	B	C	D
1	Id	Id2	Pop18to24	LessThanHighSchool
2	1400000US17031010100	17031010100	549	33
3	1400000US17031010201	17031010201	614	7.3
4	1400000US17031010202	17031010202	164	41.5
5	1400000US17031010300	17031010300	384	12.5
6	1400000US17031010400	17031010400	1439	4.1
7	1400000US17031010501	17031010501	477	28.5

ACSLessThanHighSch2005-10

9 Save the spreadsheet as **ACSLessThanHighSch.xlsx** to the Chapter5 folder of MyExercises and close Excel.

YOUR TURN

Import ACSLessThanHighSch2005-10 of ACSLessThanHighSch.xlsx into Chapter5.gdb, add it to your map document, join it to Tracts 2010, and create a choropleth map showing the percentage of population with no high school degree. Note that field LessThanHighSchool is already a percentage. Save your map document.

Tutorial 5-9

Exploring raster basemaps from Esri web services

While vector maps are discrete — consisting of points, lines connecting points, and polygons made up of lines — raster maps are continuous like photographs and use many of the same file formats as images on computers, including Joint Photographic Experts Group (.jpg) and tagged image file (.tif) formats. All raster maps are rectangular, consisting of rows and columns of cells known as pixels. Each pixel has an associated coordinate and an attribute value such as altitude for elevation. Raster maps do not store each pixel's location explicitly but rather store data such as the coordinates of the northwest corner of the map, cell size (assuming square pixels), and the number of rows and columns, from which a computer algorithm can calculate the coordinates of any cell. A key aspect of raster maps from your point of view is that they are very large files. So while you may store some important basemap raster files on your computer, these kinds of maps are perhaps best obtained as map services available for display on your computer but stored elsewhere.

Add a basemap layer

You can add several different basemap layers to a map document as services in addition to map layers from your computer or local area network. ArcGIS has built-in access to Esri servers for this purpose.

1 Save your map document as **Tutorial5-9.mxd** to the Chapter5 folder and remove all layers except County Subdivisions.

2 Symbolize the County Subdivisions layer with a hollow fill and a yellow outline, width 1.15.

3 Click File > Add Data > Add Basemap > Imagery with Labels > Add > Close. It takes a few moments for the map image to download to ArcMap. If your basemap does not appear, right-click County subdivisions and click Zoom To Layer. The map image that displays is from an Esri server instead of your computer's hard disk. You can now more clearly see Lake Michigan.

4 Zoom to an area in Chicago near Lake Michigan and Millennium Park.

YOUR TURN

Remove the current basemap from the table of contents and repeat step 3 for another basemap. Save your map document.

Tutorial 5-10

Downloading raster maps from the USGS

Next, you'll download an elevation raster from the US Geological Survey that can be added in ArcMap. The raster map in this exercise is a digital elevation model (DEM) that can be turned into a shaded relief map, meaning that it includes shadows from an artificial sun in the sky. In chapter 11 you will learn how to convert a DEM into a shaded relief model, as shown in the image below.

Launch the USGS National Map Viewer and zoom to Pittsburgh

1 In a web browser, go to http://viewer.nationalmap.gov/viewer/.

2 From the Standard toolbar, click the Zoom In Box button.

3 Zoom to western Pennsylvania and Pittsburgh where the three rivers meet until you see a scale of approximately 1:36,000 similar to the image on the next page. Note that the screen captures and layers might look different than what you see on your computer.

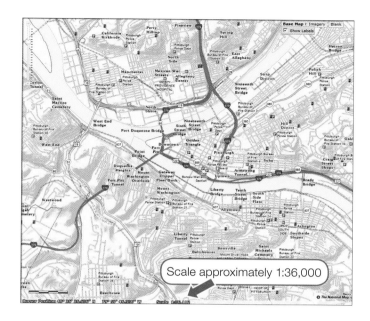

Scale approximately 1:36,000

Order data for download

1 Click the Download Data button 🔲.

2 In the "Download options" dialog box, select "Click here" to download by current map extent.

3 Select the Elevation check box and click Next.

4 Select the National Elevation Dataset (1/9 arc second) check box and click Next. This will add the raster map of your zoomed area to the checkout cart. The spatial resolution of NED 1/9 is approximately one meter per pixel, the most detailed available from the USGS.

5 In the Cart window, click Checkout, enter your e-mail address, and click Place Order. You will get a message saying that your order has been placed and to check your e-mail. Later, you'll get an e-mail message summarizing your order and telling you that a link will arrive in the future, which you can use to download the order.

Download data

At some point, you'll get an e-mail message containing a link to download the data.

1 Click the link in your e-mail to download the data. You will see the following messages in succession: "Current order status ... Adding your request to the queue. ... Extracting data ... The data extraction has completed. ... Please wait for the data to be returned."

2 Click Save File and save the downloaded file to the Chapter5 folder in MyExercises.

3 In Windows Explorer, double-click the compressed file, click the "Extract all files" button, browse to your Chapter5 folder, and click OK. Then click Extract. A folder that has a unique number is created, containing all the files that are needed to display the raster image properly in the correct projections.

4 Use Windows Explorer to view these files. Close Windows Explorer when you are finished.

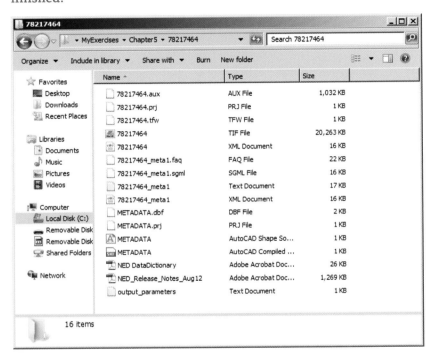

Add the raster image to the map

1 In ArcMap, create a new blank map and save it as **Tutorial5-10.mxd** to the Chapter5 folder.

2 Click the Add Data button, browse to Data, Pittsburgh folders, and City.gdb, click Neighborhoods and Add. Symbolize neighborhoods with a hollow fill.

3 Click the Add data button, navigate to MyExercises, Chapter5, and the folder with extracted raster files, click the .TIF file, click Add, and Yes to build raster pyramids. The DEM raster image will automatically be classified and added to the table of contents.

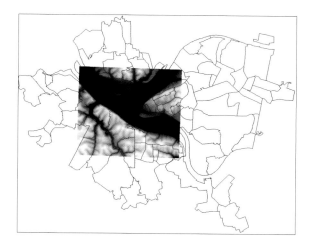

5-1
5-2
5-3
5-4
5-5
5-6
5-7
5-8
5-9
5-10
5-11
A5-1
A5-2

YOUR TURN

Download the National Land Cover Database 2006 Land Cover from the National Map Viewer and add it to your map document. You will learn more about land cover definitions and how to classify them with unique values in chapter 11. Save your map document.

Tutorial 5-11

Exploring sources of GIS data from government websites

There are many sites to download GIS data from, including local, state, and federal websites. In this exercise, you will download data from an official website of the US government, nationalatlas.gov. Data from the National Atlas can also be found in links from federal websites such as Geo.Data.gov and Geoportal.org. Agencies that contribute geospatial data to these websites include the Department of Agriculture (USDA), the Department of Commerce (DOC), the National Oceanic and Atmospheric Administration (NOAA), the US Census Bureau (CENSUS), the Department of the Interior (DOI), the US Geological Survey (USGS), the Environmental Protection Agency (EPA), and the National Aeronautics and Space Administration (NASA). Many other agencies provide other data on nationalatlas.gov and Data.gov.

Download data from nationalatlas.gov

1 Open Tutorial5-11.mxd from the Maps folder and save the map document to the Chapter5 folder.

2 Click Bookmarks > 48 Contiguous States.

3 Open your web browser and go to http://nationalatlas.gov.

4 In the search box, type **africanized bees**, press ENTER, and click Spread of Africanized Honey Bees in the United States. The direct URL to the map layer information for Africanized (killer) bees is http://nationalatlas.gov/mld/afrbeep.html

5 From the right panel under Data Download, click Invasive Species - Africanized Honey Bees.

6 Click Shapefile : afrbeep020.tar.gz, save the file to the Chapter5 folder, and uncompress the files there. The downloaded compressed file needs special software to extract the shapefile. The already extracted files, if needed, can be found in the DataGovShapefiles in the Chapter5 folder of FinishedExercises.

Import data and assign an x,y coordinate system

Sometimes downloaded GIS layers do not include a projection file with the x,y coordinate system defined. In this exercise you will import the shapefile into a file geodatabase and assign the x,y coordinate system.

1 Open Catalog, navigate to Chapter5.gdb, import the shapefile afrbeep020.shp and rename the layer **AfricanizedBees**.

2 In Chapter5.gdb, right-click AfricanizedBees and click Properties.

3 Click the XY Coordinate System tab, expand Geographic Coordinate Systems, expand North America, click NAD 1983 > OK.

4 Click the Add Data button, navigate to Chapter5.gdb, and add the AfricanizedBees layer.

Display layer using classification by year

1 In the table of contents, right-click layer AfricanizedBees, click Properties, and the Symbology tab.

2 Click Quantities > Graduated colors > YEAR as the fields value.

3 Click the Classify button, and in the Classification window click the Exclusion button.

4 Create the query "YEAR" = 0. This shows only the year that Africanized Bees began populating US states.

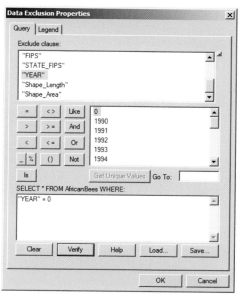

5 Click OK > OK. Click YEAR as the fields value again. This will remove years 0 and 1 from the classification ramp.

6 Change the number of classes to 4 and use the following colors:

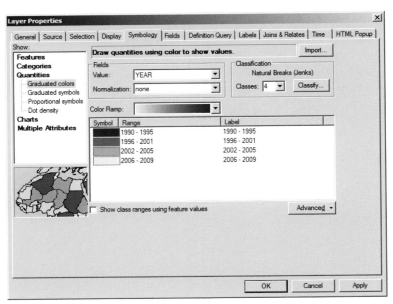

7 Click OK and move the layer just below USStates. You can now clearly see the years that Africanized bees populated US states and counties.

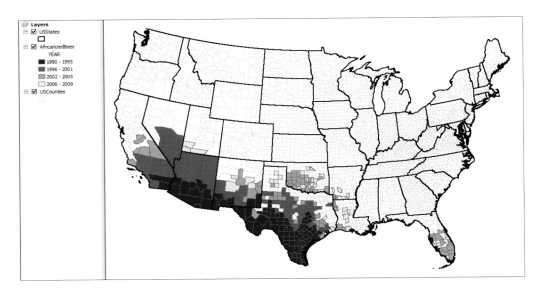

8 Save your map document.

YOUR TURN

Download two other GIS layers from the National Atlas, United States Tornado Touchdown Points 1950-2008 (http://nationalatlas.gov/mld/tornadx.html) and Bailey's Ecoregions and Subregions of the United States, Puerto Rico, and the U.S. Virgin Islands (http://nationalatlas.gov/mld/ecoregp.html). Import these shapefiles into Chapter5.gdb and assign the x,y coordinate system GCS NAD 1983. Classify ecoregions using unique values and the field DOMAIN. Create a definition query showing tornados with injuries greater than 500 with graduated symbols using the INJ field. Label tornados using the YEAR field. Save your map document and close ArcMap.

5-1
5-2
5-3
5-4
5-5
5-6
5-7
5-8
5-9
5-10
5-11
A5-1
A5-2

Assignment 5-1

Compare heating fuel types by US counties

The US Census collects data about the fuel used to heat houses in the United States. In this assignment, you explore where different fuel types are used throughout the country. To do so you download the American Community Survey data from 5-year estimates at the county level, clean and join two data tables together, and create a layout with two choropleth maps comparing fuel heating types in the country.

Get set up

- Rename the folder \EsriPress\GIST1\MyAssignments\Chapter5\Assignment5-1YourName\ to your name or student ID. Store all files that you produce for this assignment in this folder.

- Create a new map document called **Assignment5-1YourName.mxd** with relative paths.

Download and process data

Download American Community Survey data

- (http://factfinder2.census.gov). Click Topics > Housing > Physical Characteristic > Heating Fuel. Click Geographies > Geographic type County > All Counties within United States and add this to your selection. Use the search box to find tables B25040, select and download table B25040 (HOUSE HEATING FUEL) for 2010 5-year estimates and save it to your Assignment5-1YourName folder.

- In Excel, open ACS_10_5YR_B25040_with_ann.csv to clean up this data. The only attributes that you need are Id or Id2 and the heating fuel type fields. Rename these with logical names (for example, **TOTAL, UTIL_GAS, TANK_LP_GAS, ELECTRICITY, OIL_KEROSENE, COAL_COKE, WOOD, SOLAR, OTHER, NO_FUEL**). Prepare the Id or Id2 field to join to the GEOID10 fields in the USCounties feature class. Save the workbook as **HeatingFuel2010.xlsx**. Import the worksheet of the workbook into your file geodatabase as **CountyHeatingFuel2010**. Join this table to the USCounties attribute table.

Build the map

Create a new file geodatabase called **Assignment5-1YourName.gdb**. Import the following data into it:

- \EsriPress\GIST1\Data\UnitedStates.gdb\USCounties—polygon features of US counties: An attribute, GEOID10 is the county geocode in this table that you will join your downloaded table to. It has the text data type.

- \EsriPress\GIST1\Data\UnitedStates.gdb\USStates—polygon features of US States.

- You will import the worksheet of HeatingFuel2010.xlsx into your file geodatabase as **CountyHeatingFuel2010**.

Requirements

- Create a layout that compares two attributes for alternative fuel types, SOLAR and WOOD.
- Use two data frames and two separate choropleth maps, one for each new attribute.
- Rename both data frames and map layers.
- Show USStates as a hollow fill, black outline 1.5, and label with a white halo using state abbreviations.
- Use the same projected coordinate system for both layouts, Continental > North America > USA_Contiguous_Albers_Equal_Area_Conic.
- Zoom to the 48 contiguous states at the same map scale.
- Use the map layer and table in your file geodatabase (and not the shapefile or Excel table).
- Use separate quantile classifications with no decimal places, and colors that you think are effective for each map.
- List the data source and the date that census data was collected in your layout.
- Click File > Export Map and save your map as **Assignment5-1YourName.jpg** with 150 dpi resolution.

Assignment 5-2

Create a map of Maricopa County, Arizona, voting districts, schools, and voting-age population using downloaded data and a web service image

Maricopa County, Arizona is one of the nation's fastest-growing areas with 3.2 million residents and 1.5 million registered voters. Suppose that Maricopa County officials plan to use GIS to ensure accurate voting boundaries, maintain voter lists, locate polling places, plan voting precincts, recruit poll workers, and deliver voting supplies. Trained analysts will use the GIS, so it needs much relevant detail.

In this assignment you focus on skills needed to download and prepare data from websites to use GIS. You download voting districts, streets, and census block groups for the purpose of building an interactive GIS to be used in selecting schools for use as polling sites. You clean the census data, join it to the block group features, and use it to display the spatial distribution of the voting-age population. Of course, the number of registered voters by block and estimates of voter turnout would be better than voting-age population, but it's not available. An x,y table that you download shows schools that are open and their locations.

Get set up

- Rename the folder \EsriPress\GIST1\MyAssignments\Chapter5\Assignment5-2YourName\ to your name or student ID. Store all files that you produce for this assignment in this folder.
- Create a new map document called **Assignment5-2YourName.mxd** with relative paths.
- Create a new file geodatabase called **Assignment5-2YourName.gdb**. Import the spatial data and tables you will download and process into it.

Download data

Download the following 2010 TIGER shapefiles for Maricopa County, Arizona:

- Block Groups—tl_2010_04013_bg10.shp (import as **MaricopaBlockGroups**)
- Voting Districts—tl_tl_2010_04013_vtd10.shp (import as **MaricopaVotingDistricts**)
- All lines—tl_2010_04013_edges.shp (process and import as instructed below)

From http://factfinder2.census.gov, download SF 1 2010 population by sex data for Maricopa County, Arizona block groups.

- Click the Geographies link, Name tab, and Geographic Type Block Group.
- Click the plus (+) sign beside Within State, click Arizona, click Within County, Maricopa. Select the check box for All Block Groups within Maricopa County, Arizona, and Add.

- Close the Select Geographies window and scroll the pages until you find the table P12 (Sex by Age 2010 SF1 100% Data). Download and extract the file **DEC_10_SF1_P12.csv**. Be patient. This may take a while.

From http://nces.ed.gov/ccd/ (National Center for Education Statistics), download school data for Maricopa County, Arizona. Use the following steps to download open schools in Maricopa County, Arizona:

- Click the Build a Table link, select **School** for Select Row Variable, click **2009-2010** for Select Years, click Next and I Agree.
- Under Select Columns select **Basic Information (X)**, and then click **School Name-by survey year (School) 2009-10** and **County Name (School) 2009-10**.
- Under Select Columns select **Contact Information**, and then click **Location Address (School) 2009-10**, **Location City (School) 2009-10**, and **Location ZIP (School) 2009-10**.
- Under Select Columns select **School/District Classification Information**, and then click **School Type-most recent year (School) Same All Years**, **Latitude (School) 2009-10**, and **Longitude (School) 2009-10**.
- Under Select Columns select Total Enrollment, and then click **Total Students (School) 2009-10**.
- Click Next.
- Under Row Variable, select **AZ - Arizona**. Under Other Filters, select **County Name (School)**.
- In the pop-up window called Column Filter Criteria, select **2009-10** for Select Years, type **Maricopa** for Column contains, and click Save Filter.
- You'll get a large red X with County Name (School) [2009-10] contains Maricopa, but that's in case you want to delete the filter. So just leave that alone and don't click it.
- Click View Table, click the Excel link > Download Excel File, save **NCES_Report_12345.xls** Note that the report name will vary.
- Click Close.

Process the data

The shapefile, tl_2010_04013_edges.shp, has all lines used to build census blocks including lines from streets, water features, railroads, and so forth. You need to extract streets from the shapefile, using its FEATCAT attribute, to create a new feature class, MaricopaStreets. Streets have the FEATCAT code value S, while railroad tracks have R, water (hydrology) features have H, and so forth.

- Add tl_2010_04013_edges features to a new map document. Use Selection, Select By Attributes, to build the query "FEATCAT" = 'S'. In the table of contents, right-click tl_2010_04013_edges, click Data > Export Data, change the Save as type to File and Personal Geodatabase feature classes, and save the selected lines as **MaricopaStreets** in Assignment5-2YourName.gdb.

5-1
5-2
5-3
5-4
5-5
5-6
5-7
5-8
5-9
5-10
5-11
A5-1
A5-2

- Open Excel and then open dc_dec DEC_10_SF1_P12.csv to clean up this data. The only attributes that you need are **Id** or **Id2** and **AGE18UP**, which you will have to create and compute from the individual age fields. The voting age in Arizona is 18. You can compute the AGE18UP column in Excel or in ArcGIS using the field calculator. If you compute in Excel be sure that the column contains the actual data and not an expression. Save the spreadsheet as **POP18UP.xlsx**. Import the worksheet of the spreadsheet into your file geodatabase as a table called **POP18UP**.

- Open NCES_Report.xls in Excel and ignore the warning message on opening. Clean the data by deleting extraneous rows and columns and renaming columns to reasonably short names without embedded blanks (use Latitude and Longitude for those two names).

- Click File, Save As, change the Save as type to CSV (MS-DOS), type the File name as **MaricopaSchools.csv**. Note that saving the data as a CSV file gets rid of some troublesome text formatting of Latitude and Longitude so that ArcMap will import the columns as numbers.

- Import MaricopaSchools.csv into your file geodatabase as a table called **MaricopaSchools**

Requirements

- Add the census features to your map document and symbolize voting districts as a hollow fill, black outline, width 1.50 and streets as medium gray (40%).

- Join the table POP18UP to the block group layer and symbolize as a choropleth map, which represents the demand for voter registration facilities. Use a logical mathematical progression.

- Add MaricopaSchools as an x,y table using GCS_North_American_1983 (NAD 1983.prj) coordinates.

- Create a definition query so that only open schools are included (assuming enrollment over zero implies that a school is open) and export these points as a feature class called **MaricopaSchoolsOpen** in your file geodatabase.

- Symbolize schools using size-graduated point markers for total number of students (enrollment). ***Hint:*** An equal interval scale works well here.

- Use visible scale ranges to display detailed layers (streets and schools) when zoomed in below a scale of 1:50,000. Display block groups when zoomed in below a scale of 1:100,000. Display voting districts at all scales. Label voting districts and schools, but not streets or block groups. This provides a tool for analyzing potential voting places.

- Zoom to Tempe voting district 10, and then zoom out to a scale of 1:20,000. Create a bookmark called **SampleTempeVotingDistricts**.

- From the Esri web service, add basemap imagery with labels. Turn off the block group layer and export your zoomed bookmark area as an image called **Assignment5-2YourName.jpg**.

Part II
Working with spatial data

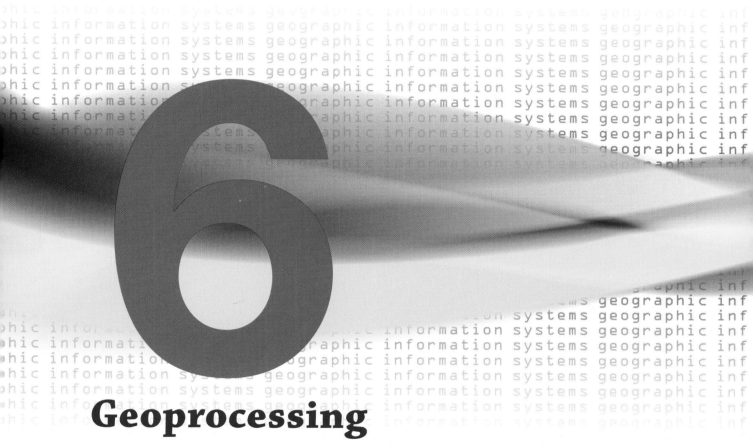

Geoprocessing

Generally you will need to use geoprocessing tools to build study areas in a GIS. In this chapter, you learn how to extract a subset of spatial features from a map using attribute or spatial queries, aggregate polygons to larger polygons, and append two or more layers into a single layer. Often it is necessary to string several such geoprocessing tools and steps together to build the desired product, so you also learn how to create macros using the ArcGIS ModelBuilder application to bundle two or more steps together into a single package.

Learning objectives

- *Use attribute and spatial queries to extract features*
- *Clip features*
- *Dissolve features*
- *Merge features*
- *Intersect layers*
- *Union layers*
- *Automate geoprocessing with ModelBuilder*

Tutorial 6-1

Extracting features for a study area

New York City is the most populated US city with more than 8 million people; it has a population density of more than 27,000 per square mile. With 195 neighborhoods in five boroughs, it is difficult to analyze or visualize the entire city at once. If city officials want to study one neighborhood in detail this can be done using ArcMap geoprocessing tools. In this tutorial you use two methods—select by attributes and select by location—to create study area features for one neighborhood.

Open a map document

1   Start ArcMap and open Tutorial6-1.mxd from the Maps folder. The map document opens showing a map zoomed to Manhattan, showing surrounding New York City metropolitan area boroughs, neighborhoods, facilities, and Manhattan streets.

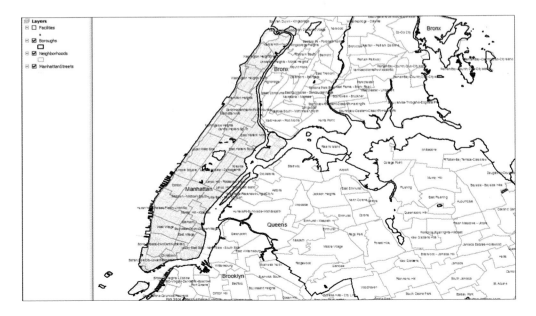

2 Click Bookmarks > Manhattan.

3 Save the map document to the Chapter6 folder of MyExercises.

Use Select By Attributes to select features

Here you use the ArcMap Select By Attributes tool to create a study area for the Lower East Side neighborhood extracted from the Neighborhoods layer.

1 On the Menu bar, click Selection > Select By Attributes.

2 Select Neighborhoods for the Layer.

3 In the Fields box, double-click "Name".

4 Click the = button.

5 Click the Get Unique Values button. Then, in the Unique Values box, double-click 'Lower East Side'.

6 Click Apply and OK.

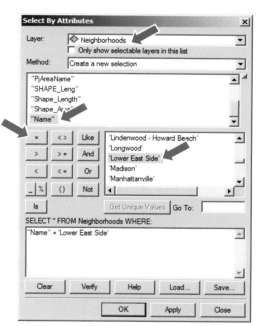

Zoom to and export selected features

1 Click Selection > Zoom to Selected Features. ArcMap zooms to the Lower East Side neighborhood in the Manhattan borough.

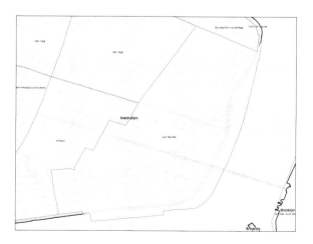

2 In the table of contents, right-click the Neighborhoods layer, click Data > Export Data.

3 Save the output feature class as **LowerEastSide** in Chapter6.gdb to the Chapter6 folder of MyExercises.

4 Click OK > Yes to add the layer to the map.

5 Symbolize the LowerEastSide layer as a hollow fill, black outline, width 2.

6 Remove the Boroughs and Neighborhoods layers. Your map now contains a new feature class containing only the Lower East Side neighborhood boundary.

Use Select By Location to select features

In the following steps, you use the Select By Location tool to select Manhattan streets that intersect the Lower East Side neighborhood only. After selecting the streets, you will create a new line feature class from them.

1 Click Selection > Select By Location.

2 Make selections as shown in the image. This selects only streets in the Lower East Side.

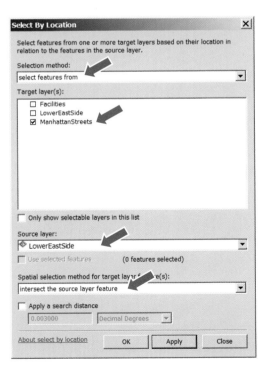

6-1
6-2
6-3
6-4
6-5
6-6
6-7
A6-1
A6-2
A6-3

3 Click Apply, Close.

4 In the table of contents, right-click the Manhattan Streets layer, click Data > Export Data.

5 Save the output feature class as **LowerEastSideStreets** in Chapter6.gdb and add the layer to your map document..

6 Symbolize the LowerEastSideStreets layer as light gray lines and zoom to this layer.

7 In the table of contents, move LowerEastSideStreets below LowerEastSide and remove the Manhattan Streets layer. You now have a study area for the Lower East Side. Using Select By Location produces streets that "dangle" past the neighborhood outline. In tutorial 6-2 you use Clip to create streets that stop at the neighborhood outline.

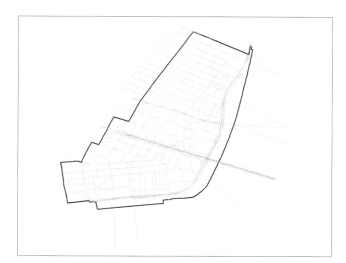

YOUR TURN

Use Select By Location to create a new feature class of LowerEastSideFacilities in Chapter6.gdb. Use "are within the source layer feature" as the selection method. Remove the original Facilities layer and symbolize the new layer using Circle 2, Mars Red, size 7. Save your map document.

Tutorial 6-2

Clipping features

Next, you will use the Clip geoprocessing tool to cleanly "cut off" the Lower East Side street segments using the Lower East Side feature class. Once this is done, the streets will have no dangling lines. Note that for geocoding tabular address data with streets, you should use the streets version with dangles because ArcMap interpolates house numbers using the starting and ending house numbers of street segments. Streets that ArcMap clips will have the original starting and ending house numbers but shortened lengths. This will introduce location errors beyond those inherent in the approximate streets. Use the clipped streets for display purposes only in a study area.

Clip streets

1 Save your map document as **Tutorial6-2.mxd** to the Chapter6 folder.

2 Click Geoprocessing > Clip.

3 In the Clip window, select LowerEastSideStreets as Input Features.

4 Select LowerEastSide for Clip Features.

5 Save the Output Feature Class as **LowerEastSideStreetsClipped** in Chapter6.gdb.

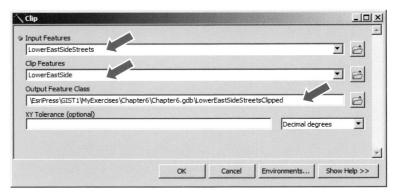

6 Click OK and wait while the streets are clipped.

7 Turn on the LowerEastSideStreetsClipped layer and turn off the LowerEastSideStreets layer. As seen in the map on the following page, the streets in the LowerEastSideClipped layer do not cross the Lower East Side neighborhood boundary.

8 Save your map document.

Tutorial 6-3

Dissolving features

You can create administrative or other types of boundaries by dissolving polygons in a feature class that share common attribute values. For each group of original polygons, dissolving retains the outer boundary lines of the group but erases interior lines. The New York City Fire Department, like most fire departments around the world, is organized in a quasi-military fashion with companies, battalions, and divisions. In this tutorial you use the Dissolve tool to dissolve New York City fire company polygons to create fire battalions and fire divisions.

Open a map document

1 Open Tutorial6-3.mxd from the Maps folder. Tutorial6-3.mxd contains a map of the New York City fire companies and boroughs.

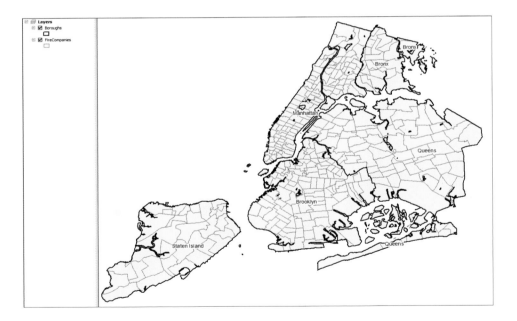

2 Save the map document to the Chapter6 folder.

Examine the dissolve attribute

1 In the table of contents, right-click FireCompanies and click Open Attribute Table. The FireBN and FireDiv attributes are used to aggregate into fire battalion and fire division polygons. The field Pop2010 shows the population in each fire company and can be aggregated during the dissolve process.

OBJECTID *	Shape *	FireCoType	FireCoNum	FireBN	FireDiv	Shape_Length	Shape_Area	Pop2010
1	Polygon	E	153	21	8	26830.204525	29213947.52536	22221
2	Polygon	E	206	28	11	30274.034187	15192741.33212	3338
3	Polygon	E	214	37	15	12069.312309	8852559.675643	17747
4	Polygon	E	216	35	11	13253.213817	9463049.329291	17847
5	Polygon	E	217	57	11	17584.543354	9738320.205785	21123
6	Polygon	E	222	37	15	14060.23304	10061437.89181	18536

FireCompanies

2 Close the table.

Dissolve fire battalions

1 Click Geoprocessing > Dissolve.

2 In the Dissolve window, select FireCompanies as the Input Features.

3 Save the Output Feature Class as **FireBattalions** in Chapter6.gdb.

4 Select FireBN as the Dissolve field.

5 Select POP2010 for the Statistics Field(s).

6 Click the Statistic Type drop-down and choose SUM. This is an optional setting. When the dissolve runs, ArcGIS sums the values in the POP2010 field for each group of polygons with the same FireBN value. In other words, it aggregates the population up to the new polygon feature of fire battalions.

7 Verify your selections with the image:

8 Click OK and wait while the polygons dissolve.

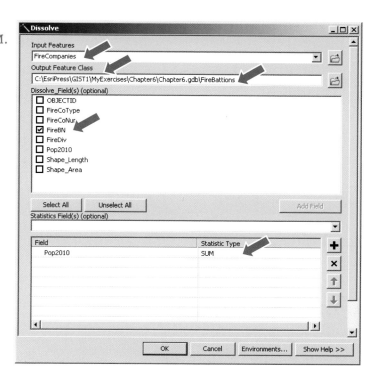

6-1
6-2
6-3
6-4
6-5
6-6
6-7
A6-1
A6-2
A6-3

9 Symbolize the new layer with a hollow fill, an Ultra Blue outline color, and an outline width of 1.15.

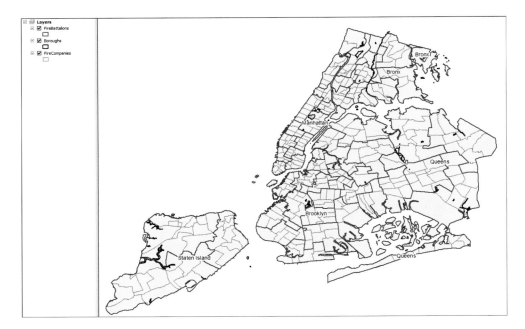

10 Zoom to Lower Manhattan and use the Identify tool to view the attribute information for fire battalion number 4 in the Lower East Side neighborhood. The SUM_POP2010 value is the aggregate of the dissolve process derived from the POP2010 values of the FireCompanies layer.

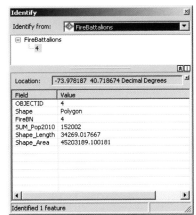

YOUR TURN

Use the Dissolve tool and field FireDiv to create fire divisions. Sum field POP2010 while dissolving. Symbolize the new layer with no fill, a Tuscan Red outline color, and an outline width of 1.15. Save your map document.

6-1
6-2
6-3
6-4
6-5
6-6
6-7
A6-1
A6-2
A6-3

Tutorial 6-4

Merging features

Sometimes it is necessary to merge two or more separate but adjacent layers into a single layer. For example, you may want to build a water layer for an environmental study that includes water layers from several adjacent counties. New York City is made up of five boroughs, each of which is also a county. The Bronx is also Bronx County, Brooklyn is Kings County, Manhattan is New York County, Queens is Queens County, and Staten Island is Richmond County. Here you will merge water layers for New York City's counties.

Open a map document

1 Open Tutorial6-4.mxd from the Maps folder. Tutorial6-4.mxd contains a map of the New York City area counties' water features, shown as separate layers.

2 Save the map document to the Chapter6 folder.

Merge several feature layers into one feature class

1 Click Geoprocessing > Merge.

2 In the Merge window, select all five layers for the New York area water polygons for Input Datasets.

3 For the Output Feature Class save **NYCWater** in Chapter6.gdb.

4 Verify that the Merge settings match the image.

5 Click OK.

6 In the table of contents, remove the individual county water layers. The NYCWater layer now contains water boundaries for all five counties around New York City.

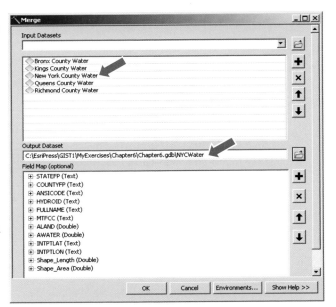

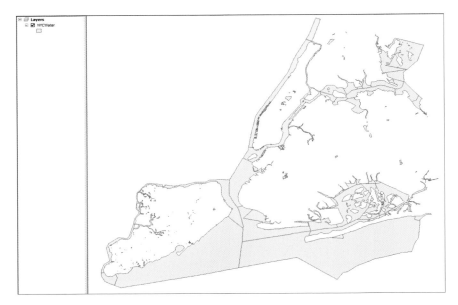

7 Save your map document.

YOUR TURN

From the NYC.gdb file geodatabase in the Data folder add the following layers: BronxWaterfrontParks, BrooklynWaterfrontParks, ManhattanWaterfrontParks, QueensWaterfrontParks, and StatenIslandWaterfrontParks. Use the Merge tool to create one feature class called NYCWaterfrontParks in Chapter6.gdb. Save your map document.

Tutorial 6-5

Intersecting layers

The Intersect tool creates a new feature class combining all the features and attributes of two input, overlaying feature classes. For example, an emergency preparedness official might like to know the name of the fire company that each street crosses over (or intersects). ArcGIS can provide such information using the Intersect tool. Intersect excludes any parts of the two or more input layers that do not overlay each other.

Open a map document

1 Open Tutorial6-5.mxd from the Maps folder. The Tutorial6-5.mxd file opens with a map of the New York City fire companies and Manhattan streets.

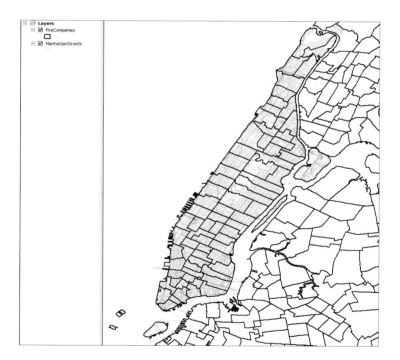

2 Click Bookmarks > Manhattan.

3 Save the map document to the Chapter6 folder.

Open tables

Next you open the table of both feature classes to see their attributes.

1 From the table of contents, right-click the ManhattanStreets layer, click Open Attribute Table, and scroll to the right. There is no data about the fire companies in this file.

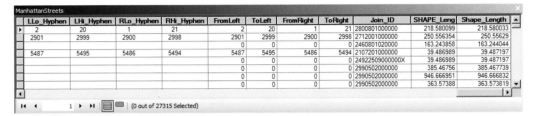

ManhattanStreets

LLo_Hyphen	LHi_Hyphen	RLo_Hyphen	RHi_Hyphen	FromLeft	ToLeft	FromRight	ToRight	Join_ID	SHAPE_Leng	Shape_Length
2	20	1	21	2	20	1	21	2800801000000	218.580099	218.580033
2901	2999	2900	2998	2901	2999	2900	2998	2712001000000	250.556354	250.55629
				0	0	0	0	2460801020000	163.243858	163.244044
5487	5495	5486	5494	5487	5495	5486	5494	2107201000000	39.486989	39.487197
				0	0	0	0	24922509000000X	39.486989	39.487197
				0	0	0	0	2990502000000	385.46756	385.467739
				0	0	0	0	2990502000000	946.666951	946.666832
				0	0	0	0	2990502000000	363.57388	363.573819

1 ▶ ▶| (0 out of 27315 Selected)

2 Open the FireCompanies table. Examine the attributes of this table. Fields of interest are the fire company, battalion, and division numbers and the type of fire company; for example, ladder company (L), engine company (E), or fire squad (Q).

FireCompanies

OBJECTID	Shape	FireCo Type	FireCoNum	FireBN	FireDiv	Pop2010	Shape_Length	Shape_Area
1	Polygon	E	153	21	8	22221	26830.204525	29213947.525366
2	Polygon	E	206	28	11	3338	30274.034187	15192741.332126
3	Polygon	E	214	37	15	17747	12069.312309	8852559.675643
4	Polygon	E	216	35	11	17847	13253.213817	9463049.329291
5	Polygon	E	217	57	11	21123	17584.543354	9738320.205785
6	Polygon	E	222	37	15	18536	14060.23304	10061437.891819
7	Polygon	E	227	44	15	9861	12410.32117	5331327.239706
8	Polygon	E	247	40	8	40804	24622.662513	21978391.299848
9	Polygon	E	248	41	15	27383	21875.193282	19248897.323501

1 ▶ ▶| (0 out of 348 Selected)

3 Close both tables.

Intersect features layers

1 Click Geoprocessing > Intersect.

2 From the Input Features drop-down list, select ManhattanStreets and Fire Companies one at a time.

3 Save the Output Feature Class as **ManhattanStreetsFire Companies** in Chapter6.gdb.

4 Select Line for Output Type.

5 Verify that the Intersect settings match the image.

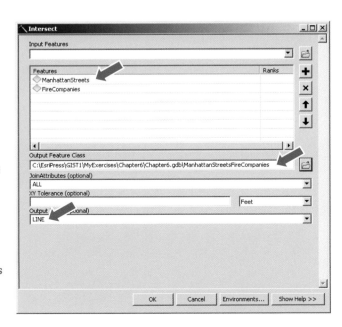

6 Click OK. The output added to your map will be Manhattan streets that intersect the fire company polygons.

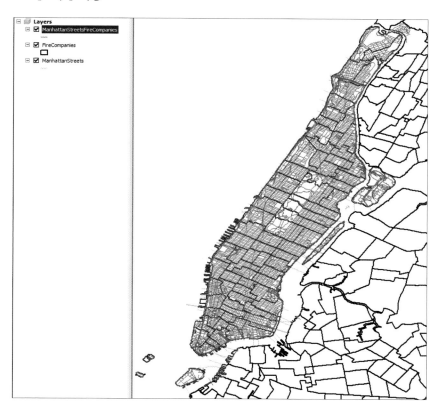

Examine the intersection table

1 In the table of contents, right-click the ManhattanStreetsFireCompanies layer and click Open Attribute Table. Each street now has data about the fire company that it intersects.

ManhattanStreetsFireCompanies

FromRight	ToRight	Join_ID	SHAPE_Leng	FID_FireCompanies	FireCoType	FireCoNum	FireBN	FireDiv	Pop2010	Shape_Length
0	0	3043401000000	417.327294	253	E	202	32	11	6541	417.327204
0	0	3631237010000	100.750667	177	L	106	28	11	17494	100.750685
0	0	1076601040000	100.476275	253	E	202	32	11	6541	100.476218
0	0	1059911010800	58.832465	311	E	35	12	3	20049	58.832551
0	0	1059911010800	63.555795	311	E	35	12	3	20049	63.555955
0	0	1043301000000	791.627747	44	E	40	9	3	27500	791.627341
0	0	1043301000000	94.674368	42	E	34	7	1	3631	94.674398
0	0	1043301000000	1960.798869	42	E	34	7	1	3631	1960.79885

I◄ ◄   1 ► ►I   [ ] [ ]   (0 out of 29627 Selected)

2 Close the table and save your map document.

Tutorial 6-6

Unioning layers

The Union tool combines the geometry and attributes of two input polygon layers to generate a new output polygon layer. In this example, you will use the Union tool to combine neighborhoods in New York City with Manhattan (New York County) census tracts. The output of the union is a new feature layer of smaller polygons, each with combined boundaries and attributes of both census tracts and neighborhoods. Union keeps all features of the input layers, even if they do not overlap.

Open a map document

1　Open Tutorial6-6.mxd from the Maps folder. The Tutorial6-6.mxd file opens showing a map of the New York City neighborhoods and Manhattan census tracts 2010.

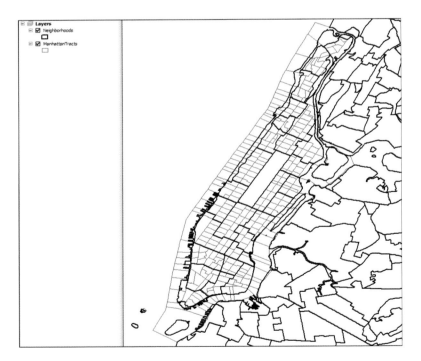

2　Click Bookmarks > Manhattan.

3　Save the map document to the Chapter6 folder.

Open tables

The union function creates a new feature class that combines the attribute tables of two feature classes and their polygon features. Next you will open the table of both feature classes to see their attributes.

1 From the table of contents, right-click the ManhattanTracts layer, click Open Attribute Table, and scroll to the right. There is no data about the neighborhoods in this file.

ManhattanTracts

AWATER10	INTPTLAT10	INTPTLON10	Id	Id2	TOT_POP	MALE	FEMALE	AGE_UNDER_1	AGE_18_64	AGE_65_UP
0	+40.8519392	-073.9342905	36061027100	36061027100	8196	4016	4180	1663	5507	1026
407508	+40.7650139	-073.9998117	36061012900	36061012900	6038	3432	2606	501	5185	352
0	+40.7752486	-073.9475520	36061014401	36061014401	4864	2161	2703	753	3245	866
1125670	+40.7523775	-074.0083669	36061009900	36061009900	1945	1002	943	192	1694	59
0	+40.7771918	-073.9521610	36061014601	36061014601	4274	1962	2312	384	3583	307
0	+40.7813494	-073.9491245	36061015400	36061015400	13749	6127	7622	1594	10395	1760
0	+40.7842784	-073.9521037	36061015801	36061015801	5650	2545	3105	894	3875	881
0	+40.7835686	-073.9475024	36061015601	36061015601	5345	2249	3096	498	4538	309

1 ▶ ▶| (0 out of 288 Selected)

2 Open the Neighborhoods table. Of course there are no tract attributes in this table.

Neighborhoods

PjAreaCode	PjAreaName	SHAPE_Leng	Shape_Length	Shape_Area	Name
MN31	Lenox Hill - Roosevelt Island	40529.675974	40529.675974	21642625.0554	Lenox Hill - Roosevelt Island
MN33	East Harlem South	18603.560972	18603.560972	16649005.3798	East Harlem South
MN12	Upper West Side	23815.836602	23815.836602	27870795.2632	Upper West Side
MN50	Stuyvesant Town-Cooper Village	10258.517828	10258.517828	5104527.21893	Stuyvesant Town-Cooper Village
MN01	Marble Hill - Inwood	39315.624604	39315.624604	18078330.4274	Marble Hill - Inwood
MN24	SoHo-Tribeca-CivcCentr-LittleItaly	20671.518956	20671.518956	23989359.053	SoHo-Tribeca-CivcCentr-LittleItaly
MN03	CentrlHarlemNorth-PoloGrounds	33385.233819	33385.233819	25591459.9506	CentrlHarlemNorth-PoloGrounds
MN15	Clinton	17878.882524	17878.882524	16433186.1111	Clinton

1 ▶ ▶| (0 out of 195 Selected)

3 Close both tables.

Select and export Manhattan neighborhoods

Before performing the union, select only Manhattan neighborhoods. Otherwise the union output will be all boroughs.

1 Click Selection > Select By Location.

2 Type or make selections as shown.

3 Click OK.

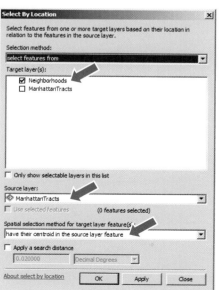

4 In the table of contents, right-click the Neighborhoods layer and click Data > Export Data.

5 Save the output feature class as **ManhattanNeighborhoods** in Chapter6.gdb.

6 Click OK, click Yes to add the layer to the map.

7 Symbolize the ManhattanNeighborhoods layer as a hollow fill, black outline, width 1.5.

8 Remove the Neighborhoods layer and zoom to lower Manhattan. Turn the ManhattanNeighborhoods layer off and on to see which polygons match the tracts boundaries. Some tracts extend beyond the neighborhood boundaries.

9 Click Bookmarks > Manhattan.

Union feature classes

1 Click Geoprocessing > Union.

2 Select ManhattanTracts and ManhattanNeighborhoods one at a time for Input Features.

3 Save the Output Feature Class as **ManhattanTractsNeighborhoods** in Chapter6.gdb.

4 Verify that the Union settings match the following:

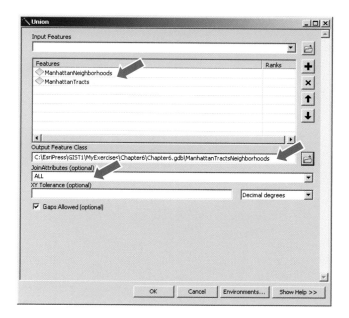

5 Click OK. The output added to your map contains many small polygons with both census tract and neighborhood data attached to each polygon. Note that the population for each new small polygon is incorrect. The Union tool merely joins the layers together and does not apportion the data across the smaller polygons.

Select tracts with neighborhoods

Many polygons from the union do not include neighborhood names because tracts and neighborhoods do not overlap. The final exercises exclude these polygons from the map document. Here you use a query to do so.

1 In the table of contents, right-click ManhattanTractsNeighborhoods > click Properties > Definition Query tab.

2 Click the Query Builder button and make selections as shown in the image.

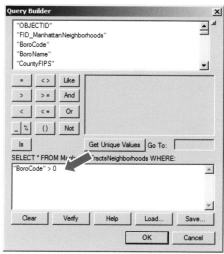

3 Click OK > OK.

The result will be only ManhattanTractsNeighborhoods with neighborhood values.

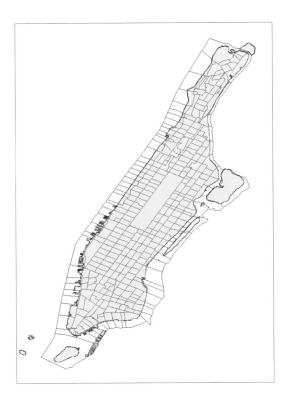

4 In the table of contents, right-click ManhattanTractsNeighborhoods and click Open Attribute Table. There are still some census tracts with no neighborhood names but these are parks/cemeteries. Notice the fields for census tracts and neighborhoods.

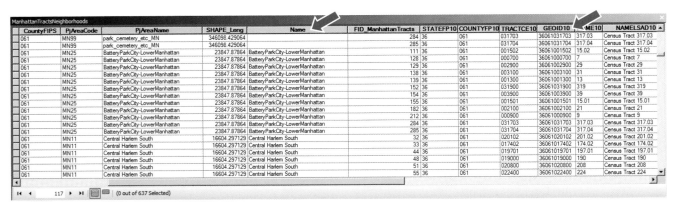

5 Close the table and save the map document.

Tutorial 6-7

Automating geoprocessing with ModelBuilder

Spatial data processing often requires several steps and geoprocessing tools to produce desired results. ModelBuilder is an application in ArcGIS for creating macros—custom programs that document and automate geoprocessing workflows. After you build a model, you can run it once, or save it and run it again using different input parameters. In this tutorial, you build a model with several steps for dissolving census tracts to make neighborhoods for a city within a county. Before you go to work using ModelBuilder, it is helpful to examine the inputs and outputs, and then the finished model that you build.

The starting map document, shown in the image, has all municipalities (cities) and census tracts in Allegheny County, Pennsylvania, as downloaded from the US Census Bureau's TIGER basemaps. In this tutorial, you create neighborhoods for Pittsburgh in the center of the map.

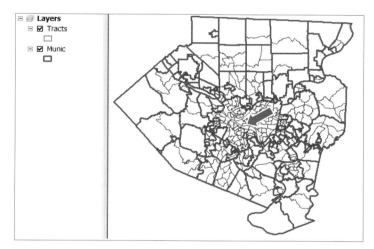

The user must supply a crosswalk table that lists the tracts that define neighborhoods in the city. In this case, each Pittsburgh neighborhood is made up of one or more tracts, as seen in the partial crosswalk table listing in this table.

OID	STFID	HOOD
0	42003220400	Allegheny Center
1	42003220100	Allegheny West
2	42003180300	Allentown
3	42003160300	Arlington
4	42003160400	Arlington Heights
5	42003202300	Banksville
6	42003050900	Bedford Dwellings
7	42003191600	Beechview
8	42003192000	Beechview
9	42003180900	Beltzhoover
10	42003080200	Bloomfield
11	42003080400	Bloomfield
12	42003080600	Bloomfield
13	42003080900	Bloomfield
14	42003090300	Bloomfield
15	42003010300	Bluff

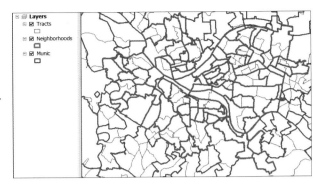

The output is the dissolved set of neighborhoods. You can see the tracts that ModelBuilder dissolved for each neighborhood as interior black lines for the red neighborhoods.

When you run the model, a form opens asking you to supply parameters—all of the elements in the model that the user needs to change for the particular run. With the model, it is possible to create dissolved polygons for any subset of a polygon basemap. See the following figure for a description of this model's parameters. This is the user interface for the model that has documentation and parameters that the user can change.

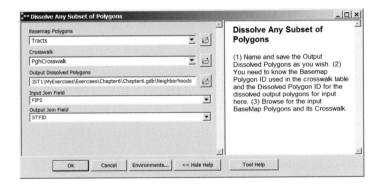

Finally, following is the workflow model diagram that you build. Earlier in this chapter you ran geoprocessing steps interactively from the ArcMap main menu. For models, however, you access the same functionality using the ArcToolbox window and tools. Each tool becomes a process (the yellow boxes seen in the model in the following figure) with blue inputs and green outputs in a model diagram.

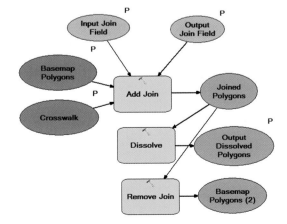

ModelBuilder shows input and output elements with black arrow lines that go from and to processes. Each element with a "P" near its upper right is a parameter. Input Join Field and Output Join Field are variables that store input parameter values for further processing (an explanation is given later in this tutorial).

The first step of the model is to join the crosswalk table to the basemap polygons. The user can supply any two consistent inputs to Add Join that have matching polygon IDs. The output, Joined Polygons, has only polygons included in the crosswalk table. In the case that you will run, the crosswalk is for Pittsburgh tracts, so only Pittsburgh tracts are output from the larger county basemap.

Next, the Dissolve process uses the crosswalk data to carry out dissolving, resulting in Output Dissolved Polygons. Finally, the model removes the join so that you can rerun the model with the same or different initial inputs. Otherwise, an error would result, indicating that the join already exists.

6-1
6-2
6-3
6-4
6-5
6-6
6-7
A6-1
A6-2
A6-3

Open a map document

1 Open Tutorial6-7.mxd from the Maps folder. The map of Allegheny County opens displaying TIGER file census tract and municipality polygons. Municipalities is just for reference, while Tracts is an input for dissolving. The other input, the crosswalk table PghCrosswalk, is also available in the map document.

2 In the table of contents, click the List by Drawing Order button.

3 Save the map document to the Chapter6 folder of the MyExercises folder.

Set geoprocessing options

1 Click Geoprocessing > Geoprocessing Options.

2 If it is not already selected, make sure that "Overwrite the outputs of geoprocessing operations" is checked. With this option on, you can rerun the model repeatedly without having to delete model outputs first, which saves time when debugging and getting your model to work properly.

3 Click OK.

Create a new model

1 Click Windows > Catalog.

2 Expand Home—MyExercises\Chapter6 in the folder/file tree.

3 Right-click Home—MyExercises\Chapter6, click New and Toolbox, and rename the new toolbox **Chapter6.tbx**.

4 Right-click Chapter6.tbx and click New, Model. ArcMap opens the Model window that you use to create your model.

Join the crosswalk table to the layer to dissolve

Next, you browse through system tools to find the Dissolve tool. When you pursue model building on your own, you will need to systematically browse through all of the tools available to get ideas and see what is possible. When you find a tool and want to learn about it, right-click it and click Help.

1 Close Catalog and click Windows > Search.

2 Click Tools from the Search window. The result is a listing and links to the ArcGIS classification of tools available for use directly or as elements in models.

3 Click the Data Management tools link. Here you see the first page of many pages with data management toolsets.

4 Scroll down and click Joins.

5 Drag the Add Join (Data Management) tool to your model. Close the Search window.

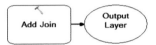

6 Double-click the Add Join process in your model and make selections using the drop-down list in each field as shown in the image. Ignore the information message symbol for the Input Join Field. Be sure to clear the Keep All Target Features (optional) check box. With this option off, the only features kept in the output are those in the crosswalk table, which will be Pittsburgh census tracts. This saves a Clip tool step.

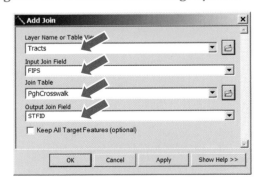

7 Click OK and resize and reposition model elements as shown in the image.

8 Click the model's Save button.

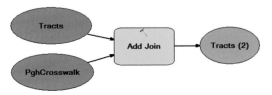

Run the partial model

ArcMap appends table names to attribute names in joined data; for example, HOOD in PghCrosswalk becomes PghCrosswalk.HOOD. Next you run the Add Join process. Then the appended attribute, PghCrosswalk.HOOD, is available in a list of attributes for you to use in configuring additional processes in subsequent steps. If you didn't run the Add Join process now, you'd have to know about and type the appended name on your own.

1 Right-click Add Join in your model, and click Run. As the process runs, a report window opens for the task and its status.

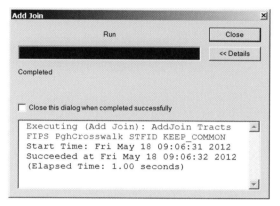

2 Click Close. ArcGIS adds shadows to the process and its output to indicate that they have been run. Note that if you make an error and have to rerun the model, first you have to click Model on the Model window's main menu and then click Validate Entire Model. This resets all processes to the unrun state.

Dissolve tracts

1 Click Windows > Search, type **dissolve** in the search text box, and press ENTER.

2 Drag the Dissolve (Data Management) tool below the Add Join process in your model. Close the Search window.

3 Click the Connect button 🔗 on the Model window's Standard toolbar, click Tracts (2) output from the Add Join process in the model, click the Dissolve process, and click Input Features in the resulting context menu.

4 Click the model's Select button ↖. You should always click this button after using another button on the Standard tool bar. Otherwise, the next time you click in the Model window, you will get an undesired action. The Select button's action is usually acceptable.

5 Double-click the Dissolve process in your model and make selections using the drop-down list in each remaining field as shown in the image on the next page (but do not click OK).

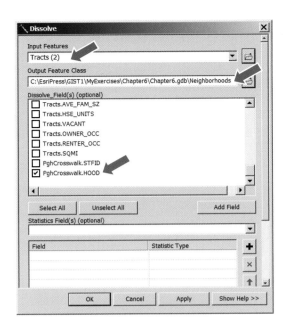

6 Select Tracts.POP2000 in the Statistics Field(s) and ignore the warning.

7 Click in the Statistic Type cell to the right of Tracts.Pop2000, click the resulting drop-down arrow, and select SUM.

8 Repeat steps 6 and 7 for two additional attributes, Tracts.WHITE and Tracts.BLACK, using SUM for both. At this point, you should have Tracts.Pop2000, Tracts.WHITE, and Tracts.BLACK in the lower panel of the Dissolve tool window, all with the SUM Statistic Type.

9 Click OK, right-click the Neighborhoods output of the Dissolve process, click Add To Display, and save your model.

10 Right-click the Dissolve process, click Run, and close the resulting window when the model has finished running.

YOUR TURN

The basic model is almost complete. The last step is to have the model remove the join in the Tracts output of the Add Join process so that the user can run the model again without manually doing so. Otherwise, the Add Join process would fail because a join already exists. Search for and add the Remove Join (Data Management) tool to the model as the last process. Use the output of Add Join, Tracts (2), as its input (double-click the Remove Join process, click "Layer Name or Table View," select the output of the Add Join process, and select Crosswalk for the Join). The Remove Join tool automatically identifies PghCrosswalk as the join to. Do not add the output of Remove Join to the display.

Run the Remove Join process. Save your model. Symbolize Output Joined Polygons (the neighborhoods) with a hollow fill and a red outline with width 3, and then move Tracts to the top of the table of contents and compare Tracts and the new Neighborhoods. You should see the output display that is at the start of this tutorial on ModelBuilder.

Reset the model so that you can run it again by clicking Model, Validate Entire Model. Run the entire model by clicking Model > Run Entire Model. Again, symbolize Output Joined Polygons with a hollow fill and red size 3 outline, move Tracts to the top of the table of contents, and compare Tracts and the new Neighborhoods. You should see the output display that is at the start of this tutorial on ModelBuilder.

Generalize element labels

6-1
6-2
6-3
6-4
6-5
6-6
6-7
A6-1
A6-2
A6-3

Your model is capable of being a general tool for dissolving any polygons. As a first step to making the model general, you change several element labels.

1 Right-click the Tracts input to Add Join element, click Rename, type **Basemap Polygons**, and click OK.

2 Similarly, change labels of other elements as shown in the image.

3 Click the model's Save button.

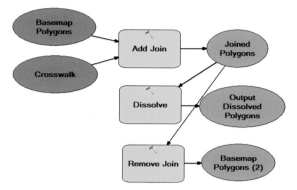

Add model parameters

Currently, your model is "hardwired" with inputs and outputs fixed in processes. Next, you will make several element parameters that users can change without modifying the model itself. Instead, users will type or make selections in a form when opening the model.

1 Right-click Basemap Polygons and click Model Parameter. A "P" appears above and to the right of the element indicating ArcGIS will ask the user to browse for an input map layer.

2 Similarly, make Crosswalk and Output Dissolved Polygons model parameters.

Add variables to the model

To be general, the Add Join process needs to get two of its inputs from the user, the Input Join Field and the Output Join Field. You can make these inputs parameters, but first you need to create variables to store them in the model.

1 Right-click the Add Join process. Click Make Variable > From Parameter > Input Join Field. ArcGIS creates the variable for you.

2 Click anywhere in the white area of the model to deselect elements. Move the new variable above the top left of the Add Join process and make its element a bit wider so that its entire label displays.

3 Make Input Join Field a model parameter.

4 Repeat steps 1 through 3, except make the variable for the Output Join Field of the Add Join process. Move the new element above and to the right of the Add Join process.

Add labels for documentation

Labels can help document the model. You will add a model title and some notes about the variables.

1 If necessary, select all model elements and make some room at the top for a label.

2 Right-click the white area at the top, click Create Label, double-click the resulting Label, and type **Model to Dissolve a Subset of Basemap Polygons**.

3 Right-click the new label, click Display Properties, click in the cell to the right of Font, click the resulting builder button, change the font to Bold size 14, click OK, and close the Display Properties window.

4 Right-click the Input Join Field element, click Create Label, click anywhere in the model white area, drag the new label above and to the left of the element, double-click it, and type **You must examine the Basemap Polygons attribute table and Crosswalk table and note their field names that share the same polygon IDs or names for use as parameter values on opening the model**.

5 Right-click the new label, click Display Properties, click the cell to the right of Text Justification, click Left instead of Center, and close the Properties window.

6 Break this label up into several lines by placing your cursor after words to create new lines, pressing the SHIFT key, and pressing ENTER.

7 Click the model's Save button.

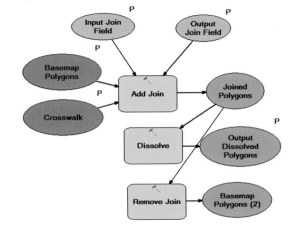

Model to Dissolve a Subset of Basemap Polygons

You must examine the Basemap attribute table and Crosswalk table and note their field names that share the same polygon IDs or names for use as parameter values when opening the model.

Add a model name and description for documentation

You can add help documentation to the form that will open on running the model, in model properties.

1 Click Model on the model's main menu, and click Model Properties.

2 Type **DissolvePolygons** for Name (no spaces are allowed) and **Dissolve Any Subset of Polygons** for Label.

3 Type as much as you want of the following help message in the Description text box: **Dissolve any subset of basemap polygons given a crosswalk table for the subset.**

    1. **Name and save the Output Dissolved Polygons as you wish.**

    2. **You need to know the Basemap Polygon ID used in the crosswalk table and the Dissolved Polygon ID for the dissolved output polygons for input here.**

    3. **Browse for the input Basemap Polygons and its Crosswalk.**

4 Select the Store relative path names check box.

5 Click OK.

6 Click the model's Save button and close the Model window.

Open and run the finished model

1 Remove Neighborhoods from the table of contents in your map document.

2 In Catalog, right-click the Dissolve Any Subset of Polygons model, click Open, and close the warning at the top of the resulting form.

3 Click the Show Help button at the bottom of the form. You do not need to change any of these input parameters.

4 Click OK to run the model. The model runs, adding the dissolved neighborhoods to your map display.

5 Save your map document and close ArcMap.

6-1
6-2
6-3
6-4
6-5
6-6
6-7
A6-1
A6-2
A6-3

Assignment 6-1

Build a study region for Colorado counties

In this assignment, you build a study area for two rapidly growing counties in Colorado: Denver and Jefferson counties. You create new feature classes for an urban area study using point and polygon layers downloaded from the US Census website. Because we want to study two counties, you need to use select by location, merge, and clip to create the study area features.

Get set up

- Rename the folder \EsriPress\GIST1\MyAssignments\Chapter6\Assignment6-1YourName\ to your name or student ID. Store all files that you produce for this assignment in this folder. Create a new map document called **Assignment6-1YourName.mxd** with relative paths.
- Create a new file geodatabase called **Assignment6-1YourName.gdb**. Store new features that you produce for this assignment in this file geodatabase.

Build the map

Add the following to your map document:

- \EsriPress\GIST1\Data\UnitedStates.gdb\COCounties—polygon features of Colorado counties.
- \EsriPress\GIST1\Data\UnitedStates.gdb\COStreets1—line features of Jefferson County streets.
- \EsriPress\GIST1\Data\UnitedStates.gdb\COStreets2—line features of Denver County streets.
- \EsriPress\GIST1\Data\UnitedStates.gdb\COUrban1—US Census 2010 urban area features for Jefferson County.
- \EsriPress\GIST1\Data\UnitedStates.gdb\COUrban2—US Census 2010 urban area features for Denver County.
- \EsriPress\GIST1\Data\UnitedStates.gdb\COWater1—US Census 2010 named water features for Jefferson County.
- \EsriPress\GIST1\Data\UnitedStates.gdb\COWater2—US Census 2010 named water features for Denver County.
- \EsriPress\GIST1\Data\UnitedStates.gdb\USCities—point features of US cities.

Process the data

Perform select by location or geoprocessing operations to create the following study area features in your file geodatabase. Remove the original features when finished with the geoprocessing.

- **StudyAreaCounties**—one feature class of Jefferson and Denver counties only

- **StudyUrbanAreas**—one feature class combining urban areas for both Jefferson and Denver counties
- **StudyAreaCities**—point layer showing populations of cities in the study urban area only. Add and calculate a field for the population change between 2000 and 2007.
- **StudyAreaStreets**—one feature class combining streets also clipped to the study urban area
- **StudyAreaWater**—one feature class combining water features that intersect the study urban area

Requirements

- Label the study area counties and cities using the NAME fields with halo masks, display the cities as graduated points using the POP2007 field with a quantile classification and 5 classes, and display the water and streets as ground features. Decide how best to display the study urban area polygons. Select Wadsworth street in the study area. *Hint:* Use FE_ NAME=Wadsworth. Show it as a separate layer.
- Create an 8.5-by-11-inch map layout zoomed to the study urban area that includes a title, legend, scale bar, and other items you think necessary. Export the layout as **Assignment6-1YourName.jpg**.

Analyze the map

Create a Word document called **Assignment6-1YourName.docx**. Answer the following questions using the attribute tables and spatial queries, and insert your layout image below your answers:

1. What is the total 2007 population of the cities in the Jefferson County study areas?
2. What is the total 2007 population of the city in Denver County?
3. What city had the largest population increase between 2000 and 2007?
4. What city had the largest population decrease between 2000 and 2007?
5. Name the cities that are within two miles of Wadsworth roads in the study urban area. *Hint*: Select all streets, boulevards, parkways, etc. named Wadsworth.

Assignment 6-2

Dissolve property parcels to create a zoning map

In this assignment, you dissolve a parcel map to create a zoning map that highlights a proposed commercial development in what is now a residential area. A commercial company wants to apply for a zoning variance so that it can use the land in residential parcels with zoning code R5 (residential dwelling with five units), for a commercial purpose. The zoning department wants the map for a public hearing on the proposal for use in a PowerPoint presentation.

Get set up

- Rename the folder \EsriPress\GIST1\MyAssignments\Chapter6\Assignment6-2YourName\ to your name or student ID. Store all files that you produce for this assignment in this folder.
- Create a new map document called **Assignment6-2YourName.mxd** with relative paths and data frame map units in feet.
- Create a new file geodatabase called **Assignment6-2YourName.gdb**. Store new features that you produce for this assignment in this file geodatabase.

Build the map

Add the following coverages to your map document:

- \EsriPress\GIST1\Data\Pittsburgh\EastLiberty\EastLib—polygon coverage for the East Liberty neighborhood boundary.
- \EsriPress\GIST1\Data\Pittsburgh\EastLiberty\Parcel—polygon coverage for land parcels in the East Liberty neighborhood of Pittsburgh. Attributes include: ZON_CODE, an attribute with the following zoning code values: A = development, C = commercial, M = industrial, R = residential, S = special; TAX_AREA_A, TAX_BLDG_A, and TAX_LAND—attributes with annual property tax values.
- \EsriPress\GIST1\Data\Pittsburgh\EastLiberty\Curbs—polyline coverage that has street curbs and annotation with street names.

Requirements

- Export the Parcels coverage to your file geodatabase and add it to your map document. Remove the original Parcels coverage.
- Select parcels with ZON_CODE = R5. These are parcels that could be converted for a commercial purpose. Add these parcels as a new layer renamed **Proposed Commercial Parcels** shown with red color fill and black outline.
- Create an aggregate-level zoning code by adding a new field to the parcels attribute table that has just the first character of the full zoning code. Call the new field **ZONE** with text data type and length 1. Using the Field Calculator on the new field, click the String Option button, click the Left( ) function, type to yield **Left([ZON_CODE],1)**, and click OK. The left

function extracts the number of characters entered—1 in this case—starting on the left of the input field, ZON_CODE.

- Dissolve the parcel's features using the Dissolve function, using your new field (ZONE) as the dissolve field, and adding SUM statistics for the three tax fields (TAX_LAND_A, TAX_BLDG_A, TAX_AREA). Click in the Statistics Type cells to select SUM. Save the output feature class called **Zoning** in your file geodatabase.

- Add the new Zoning feature class to your map document, as well as the curbs arcs and curbs annotation, and East Liberty outline. Use the Unique Values option of the categories method of classification for symbolizing the ZONE field. Use muted colors with no outlines for the various ZONE values, including Lilac Dust for A (Development), Rose Quartz for C (Commercial), Gray 30% for M (Industrial), Yucca Yellow for R (Residential), and Blue Gray Dust for S (Special). Label each zoning code with its full name. Add zoning features again as a light gray, hollow fill outline.

- Create an 8.5-by-11-inch portrait layout zoomed to scale 1:9,000 including a title, legend, map scale in feet, and other map elements that you think necessary.

- Export your map layout as **Assignment6-2YourName.jpg**.

Create a PowerPoint document

Create a PowerPoint presentation called **Assignment6-2YourName.pptx** that includes the following:

- Title slide, including your name
- Slide with a table of the summed Tax values for each zoning code
- Slide with your map layout image

Assignment 6-3

Build a model to create a fishnet map layer for a study area

The following image is the fishnet map layer that you will create with several steps saved in a ModelBuilder model. It consists of uniform, square grid cells saved in a polygon map layer, and as an option includes an additional layer of centroid points for each cell. Both the cells and centroids are useful for spatial analysis; for example, for displaying counts of structural fires by grid cell. The centroids allow you to display a second attribute using size-graduated point markers or a color ramp, while a choropleth map displays the first variable. You can create cell-level data from point data, such as residential fire incidents or crimes locations, using spatial overlay as done in chapter 2.

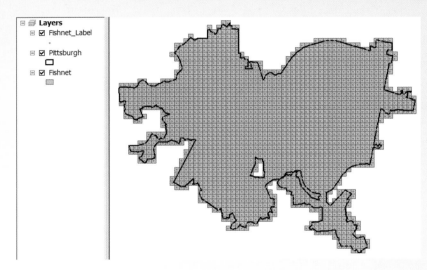

Get started

First, rename your assignment folder, create a map document, create a file geodatabase, and create a toolbox and model.

- Rename the folder \EsriPress\GIST1\MyAssignments\Chapter6\Assignment6-3YourName\ to your name or student ID. Store all files that you produce for this assignment in this folder.
- Create a new map document, **Assignment6-3YourName.mxd** with relative paths.
- Create a new file geodatabase, **Assignment6-3YourName.gdb**. Save all of the map layers and tables you create in this file geodatabase.
- Add a new toolbox called **Assignment6-3YourName** with a model called FishnetStudyArea.
- Add \EsriPress\GIST1\Data\Pittsburgh\City.gdb\Pittsburgh to your map document, symbolized with a hollow fill.

Requirements

The model diagram for the finished model that you create is in the figure. The input map layer is any polygon layer that defines the study area. It could be a boundary, such as for Pittsburgh, census tracts, or any other set of polygons. The Create Fishnet tool (see tutorial 2-7) has several parameters for constructing the grid cell map layer. It needs the cell size (while the tool has inputs for both width and height, you will almost always want square grid cells, so width will equal height), the number of rows and columns in the extent, and the extent coordinates of the output. The Create Fishnet tool has provision to import the needed map extent coordinates from the input study area polygons, so that part is easy. To calculate the number of rows and columns of the fishnet, you must have projected coordinates for the input layer (the Pittsburgh layer has projection state plane in feet, so this case meets that requirement) and you must look up its map extent coordinates in its Source properties as seen in the image as Top, Bottom, Left, and Right. You can get these extent values by right-clicking Pittsburgh in the table of contents and opening its property sheet to the Source tab. Then use the Excel workbook, \EsriPress\GIST1\Data\DataFiles\FishnetCalculations.xlsx, to compute the number of rows and columns for any cell size. For the case of 1,000-foot cell size, you need 52 rows and 64 columns.

	A	B	C	D	E	F
2						
3	Cell size	1,000	Can be changed to whatever you wish			
4						
5	**Number of rows**					
6	Top	433,417				
7	Bottom	381,929		Divide Height by cell size		
8	Height	51,487		number cells =	51.48744	
9				**rounded up =**	**52**	
10	**Number of columns**					
11	Right	1,379,786		Divide Width by cell size		
12	Left	1,315,935		number cells =	63.851	
13	Width	63,851		**rounded up =**	**64**	
14						

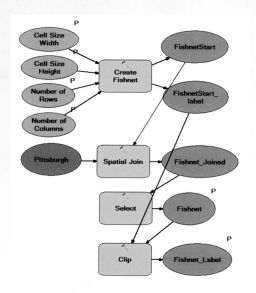

Notes on the Create Fishnet process follow:

- Create the three variables Cell Size, Number Rows, and Number Columns. Select Double for the data type of CellSize and Long for the other two variables' data type (which is integer).
- For the Create Fishnet process, name the study area polygon layer (Pittsburgh) **StudyArea** and select it for the Template Extent input.
- For the Create Fishnet's Cell Size Width and Height inputs, use the drop-down lists and select Cell Size (the name of your model variable). For the Number of Rows and Number of Columns inputs, use the drop-down lists and select your corresponding variables.
- For the Geometry, choose POLYGON.

The fishnet process creates a rectangular map layer, but what is needed is a grid cell map limited to the shape of the study area. It is best to leave grid cells whole squares, with some overhanging the study area, rather than clipping them. The model needs two steps to accomplish this. First you have to assign study area attributes to grid cells with the Spatial Join process. Then all grid cells in the interior or crossing the boundary of the study area will have non-null attributes while remaining cells will have null values. Then you can use the Select process with the criterion:

NAME > "

where there are two single quotes after the greater than sign, signifying a null value. The criterion selects all cells having non-null values.

Finally, you can clip the label points using the finished fishnet. Making the final outputs (Fishnet_Finished and Fishnet_LabelPointsFinished) parameters allows the user to rename and save them wherever desired.

7

Digitizing

Using existing map layers such as streets for guides, in this chapter you learn how to create and edit spatial data. You learn how to create and digitize new vector features and add their attribute data. Many advanced digitizing tools are available to assist you, making the work easier and more precise. You also adjust vector data spatially to make it align with a basemap layer. For example, you learn how to import and adjust a computer-aided design drawing for use in a map layer.

Learning objectives

- *Digitize polygon, line, and point features*
- *Add attribute data for vector features*
- *Learn advanced editing tools*
- *Spatially adjust features*

Tutorial 7-1

Digitizing polygon features

You create a new polygon feature class and then add features to it using heads-up digitizing (with your head up, looking at your computer screen).

Open a map document

1 Start ArcMap and open Tutorial7-1.mxd from the Maps folder. The Tutorial7-1 map document shows a map of the Middle Hill neighborhood of Pittsburgh. You use the Commercial Properties and Street Centerlines layers as references for digitizing commercial zones.

2 Save your map document to the Chapter7 folder of MyExercises.

Create a new polygon feature class

1 Open the Catalog window.

2 In the Catalog tree, browse to MyExercises, the Chapter7 folder.

3 Right-click Chapter7.gdb, click New > Feature Class.

4 In the Name field of the New Feature Class window, type **CommercialZones**.

5 For Type, select Polygon Features and click Next.

6 Expand Projected Coordinate Systems > State Plane > NAD 1983 (US Feet), click NAD 1983 StatePlane Pennsylvania South FIPS 3702 (US Feet), and click Next three times.

7 Type **ZoneNumber** as a new field, select Short Integer as the Data Type, and click Finish.

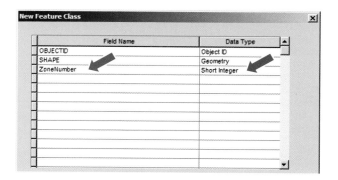

7-1
7-2
7-3
7-4
7-5
A7-1
A7-2

The result is a new polygon feature class added to Chapter7.gdb and the table of contents. While CommercialZones appears in the table of contents, of course nothing displays for it because at this point there are no features in this new map layer. Next, you digitize new features, starting with some practice polygons.

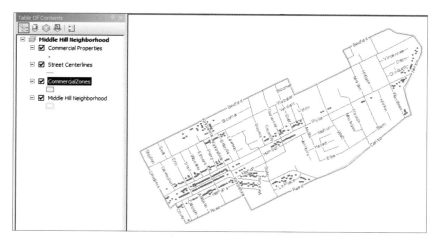

8 Close the Catalog window.

Start editing with the Editor toolbar

1 Click Customize > Toolbars > Editor. The Editor toolbar appears. You can move it or dock it anywhere in ArcMap. Dock it on top of the ArcMap window below the Standard toolbar.

2 On the Editor toolbar, click Editor > Start Editing.

3 Click CommercialZones as the layer to edit and click OK and Start editing.

4 From the Editor toolbar, click the Create Features button 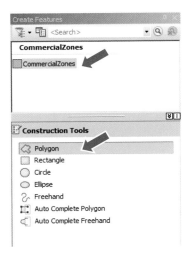 . The Create Features and Construction Tools panels appear on the right of the map. You can adjust these panels by dragging the boundary between them.

5 Click CommercialZones in the Create Features panel, and then click Polygon in the Construction Tools panel.

6 Close the Create Features panel.

Practice digitizing a polygon

1 In the table of contents, click the List By Selection button and make CommercialZones the only selectable layer.

2 On the Editor toolbar, click the Straight Segment button ✎.

3 Position the crosshair cursor anywhere on the map outside the neighborhood boundary and click to place a vertex.

4 Move your mouse and click a series of vertices one at a time to form a polygon (but not double-click!). If you click near an existing feature, you find that your new vertices snap to existing vertices of other features. You learn how to turn this behavior on and off later.

5 Double-click to place the last vertex.

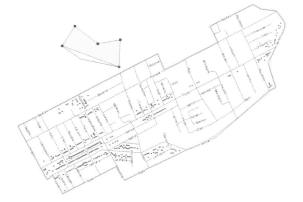

Move a polygon

1 On the Editor toolbar, click the Edit tool  .

2 Click and hold down the mouse button anywhere inside your new polygon.

3 Drag the polygon a small distance.

Delete a polygon

1 Click anywhere inside your new polygon.

2 Press the DELETE key on the keyboard.

7-1
7-2
7-3
7-4
7-5
A7-1
A7-2

> ### *YOUR TURN*
> Practice creating new polygons using the Polygon, Rectangle, Circle, and Ellipse tools from the Create Features, Construction Tools panel. Delete your practice polygons when finished.

Edit polygon vertex points

Next, you learn how to work with vertices. You move, add, and delete vertices from a polygon.

1 Click Bookmarks > Erin Street. Your map zooms to a few city blocks surrounding Erin Street.

2 Click the Create Features button and from the Construction Tools panel, click Polygon.

3 Click the Straight Segment button and draw a new polygon feature as shown in the image, snapping to street centerline intersections of Davenport, Webster, Erin, and Wylie.

4 On the Editor toolbar, click the Edit button.

5 Double-click the new polygon. Grab handles—small squares—appear on the polygon at its vertex locations.

Next, you see that you can edit the shape of a feature by moving a vertex.

6 Position the cursor over the lower right vertex point.

7 Drag the vertex to the street intersections of Wylie and Trent. The polygon's shape changes correspondingly.

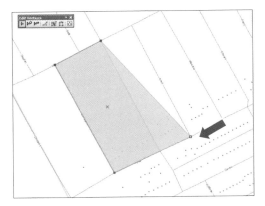

8 Drag the upper right vertex to the intersection of Webster and Trent streets.

9 Click anywhere on the map or polygon to confirm the new shape.

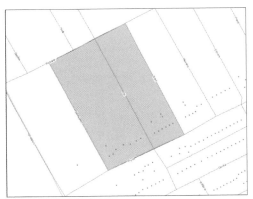

Add and move vertex points

Next, you practice editing digitized polygons and learn how to add, delete, and move vertices.

1 Double-click inside the polygon. Grab handles appear on the polygon and the small Edit Vertices toolbar appears.

2 On the Edit Vertices toolbar, click the Add Vertex button.

3 Move the mouse along Wylie Street between the two existing vertices and click. This adds a new vertex at the location of the cursor. Now you can move the new vertex to change the polygon's shape.

4 Position the cursor over the new vertex, then drag the vertex to the intersection of Wylie and Erin streets.

5 Click anywhere on the map to confirm the new shape.

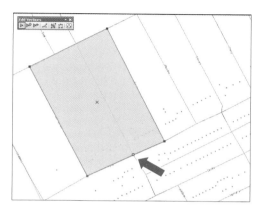

YOUR TURN

Practice adding and moving vertex points for the intersections of Webster and Erin streets and Webster and Seal streets.

Delete vertex points

1 Double-click inside the new polygon.

2 Click the Delete Vertex button.

3 Place your cursor over the vertex point between Webster and Erin and click.

4 Click anywhere on the map to confirm the new shape.

YOUR TURN

Practice changing the shape of the new polygon by moving, adding, and deleting vertices. When finished, delete the polygon.

Digitize commercial zones

1 Click Bookmarks > LaPlace. This shows the commercial block centroids in this city block. The image shown has the polygon drawn that you are about to roughly digitize.

2 In the Construction tools panel, click Polygon.

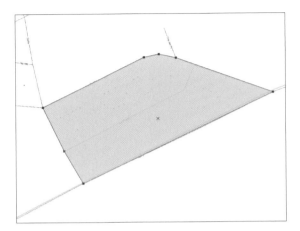

3 On the Editor toolbar, click the Straight Segment tool and digitize the LaPlace street polygon by clicking one vertex at a time and double-clicking to finish. Wherever possible, use street centerlines as a guide for digitizing your polygon.

YOUR TURN

Zoom in to a part of your new polygon and use the Add, Delete, and Move Vertex tools to refine the polygon's shape. Use the Pan tool on the Tools toolbar to move around your polygon's boundary and eventually refine all of it. You need to alternate between the Edit tool, confirming a change, and the Pan tool. Click the Full Extent button and then zoom in to a cluster of commercial points to digitize another polygon. Repeat until you have digitized all polygons seen in the following image. When you complete the final polygon, click Editor and save your edits.

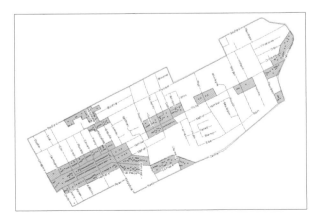

Edit feature attribute data

Now that you have digitized the commercial polygons, you assign zone numbers to them.

1 Open the CommercialZones attribute table.

2 Click in the first cell of the ZoneNumber field, type **1**, and press ENTER.

3 In sequential order, continue numbering the remaining cells in the ZoneNumber field.

4 Click Editor, Stop Editing, and Yes to save your edits.

5 Close the attribute table.

Label commercial zones

1 Turn off the Commercial Properties layer.

2 In the table of contents, right-click the CommercialZones layer and click Properties.

3 Select the Labels tab and type or make selections as shown.

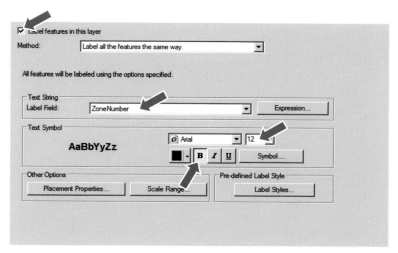

4 Click OK. Your label numbers may not match those in the following image, depending on your order of digitizing.

5 Save the map document.

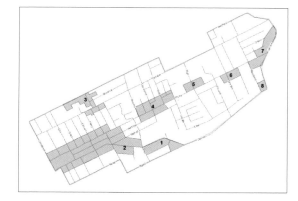

Tutorial 7-2

Digitizing line features

City planners often work with transportation experts to decide bus route and stop locations. Existing street centerlines can be used as a base layer to digitize bus route locations. In this exercise you create the bus routes that are in the Middle Hill neighborhood of Pittsburgh.

Create a line feature class for bus routes

1 Open Tutorial7-2.mxd from the Maps folder and save it to the Chapter7 folder.

2 Click Windows > Catalog.

3 In the Catalog tree, browse to Chapter7.gdb in the Chapter7 folder.

4 Right-click Chapter7.gdb, click New > Feature Class.

5 In the Name field, type **BusRoutes** and select Line Features for Type.

6 Click Next and NAD_1983_StatePlane_Pennsylvania_South_FIPS_3702_Feet as the Current coordinate system.

7 Click Next three times.

8 Type **ROUTE** as a new field, select Text as the Data Type, click Finish, and close the Catalog window. The new bus route feature class will be added to your table of contents and you are now ready to digitize the routes.

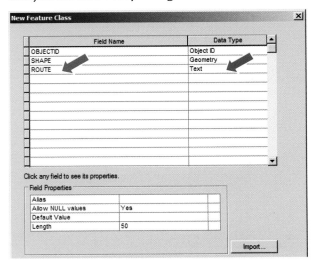

7-1
7-2
7-3
7-4
7-5
A7-1
A7-2

Symbolize the bus route features

1 In the table of contents, click BusRoutes's line symbol to open the Symbol Selector window.

2 Change the bus routes color to Ultra Blue, width 1.15.

Prepare area for digitizing and start editing

1 Click Bookmarks > Centre Avenue Route to zoom to area as shown in the image.

2 On the Editor toolbar, click Editor > Start Editing.

3 Click BusRoutes > OK.

4 On the Editor toolbar, click the Create Features button.

5 In the Create Features panel click BusRoutes > Line in the Construction Tools panel.

Set endpoint and vertex snapping

Here you turn on only endpoint and vertex snapping so you snap to the endpoint of the street segment only. This makes digitizing the bus routes faster.

1 Click Customize > Toolbars > Snapping.

2 Deselect the point and edge buttons so only the endpoint and vertex buttons are on.

Digitize by snapping to features

1   On the Editor toolbar, click the Straight Segment tool.

2   Click the endpoint of the left-most Centre Avenue street centerline to choose the route's starting point.

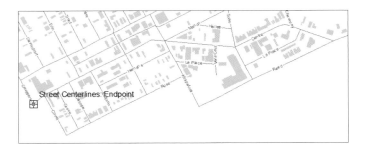

3   Move the cursor to the next street endpoint (Centre and Covet) and click.

4   Continue snapping to street intersections along the bus route shown in the image and double-click to finish the route at the intersection of Centre and Reed streets.

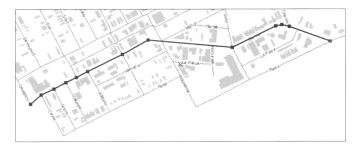

Enter bus route number

It is often easiest to enter the data for a feature as you digitize it. Here you enter the bus route number for the route that you just digitized.

1 In the table of contents, right-click BusRoutes and click Open Attribute Table.

2 Type **81A|81B|84A|84C** as the ROUTE.

3 Close the BusRoutes attribute table.

YOUR TURN

Digitize bus routes for Bedford/Erin, Webster, Wyle, and Kirkpatrick streets. Use spatial bookmarks to zoom to the route locations. Enter the route numbers and label as shown in the figure. Be sure to digitize one line for each route. When finished, stop editing and save the edits to your bus routes. Save the map document.

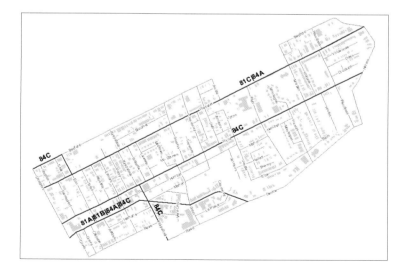

Tutorial 7-3

Digitizing point features

In addition to bus routes, point features for bus stops are needed for planning purposes. In this exercise, you digitize bus stop locations as points.

Create a point feature class for bus stops

1 Save your map document as **Tutorial7-3.mxd** to the Chapter7 folder.

2 Click Windows > Catalog.

3 In the Catalog tree, browse to Chapter7.gdb.

4 Right-click Chapter7.gdb, click New > Feature Class.

5 In the Name field, type **BusStops**.

6 Select Point Features for Type and click Next.

7 Click NAD_1983_StatePlane_Pennsylvania_South_FIPS_3702_Feet and click Next three times.

8 Type **STOP_NAME** as a new field, and select Text as the Data Type.

The BusStops point features are added to your map document and you are ready to digitize them.

9 Click Finish and close the Catalog window.

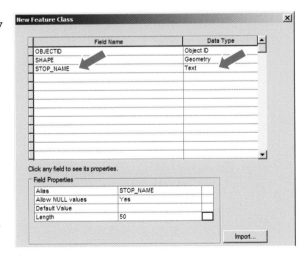

Turn snapping off

Bus routes in the Middle Hill neighborhood are offset from the street centerlines so you need to turn feature snapping off.

1 Click Customize > Toolbars > Snapping.

2 Deselect the endpoint and vertex buttons to turn all snapping options off.

Prepare the map and digitize points

1 In the table of contents, click the legend symbol for the BusStops layer. Change the symbol to Circle 2, the color to Mars Red, and the size to 8, and click OK.

2 Zoom to the Centre Avenue Route bookmark.

3 From the Editor toolbar, click Editor > Start Editing.

4 Click BusStops > OK.

5 On the Editor toolbar, click the Create Features button.

6 Click BusStops in the Create Features panel and Point in the Construction Tools panel.

7 Click to place the following approximate bus stop points along the Centre Avenue route:

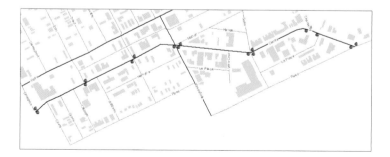

8 When finished, click Editor > Stop Editing > Yes to save the BusStop points.

7-1
7-2
7-3
7-4
7-5
A7-1
A7-2

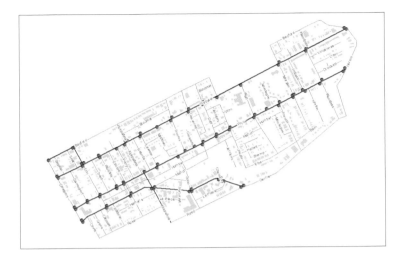

YOUR TURN

Digitize bus stops for Bedford/Erin, Webster, Wyle, and Kirkpatrick streets. Use spatial bookmarks to zoom to the route locations. Enter the fictitious bus stop names (for example Centre and Soho) for a few random stops. When finished, stop editing and save the edits to your bus stops. Add the CommercialZones polygons that you created in tutorial 7-1. Does it appear that the bus routes and stops serve the commercial zones? Save the map document.

7-1

7-2

7-3

7-4

7-5

A7-1

A7-2

Tutorial 7-4

Using advanced editing tools

There are several advanced editing tools. Here you try the Trace, Cut Polygons, Smooth, Generalize, and Rotate tools, all of which affect the shape of digitized polygons.

Create a new polygon feature class

1 Open Tutorial7-4.mxd from the Maps folder and save as **Tutorial7-4.mxd** to the Chapter7 folder.

2 Click Windows > Catalog.

3 In the Catalog tree, browse to Chapter7.gdb.

4 Right-click the Chapter7.gdb folder, click New > Feature Class.

5 In the Name field, type **CampusStudyArea** and select Polygon Features for Type.

6 Click Next and NAD_1983_StatePlane_Pennsylvania_South_FIPS_3702_Feet as the Current coordinate system.

7 Click Next three times, click Finish, and close Catalog. The new CampusStudyArea feature class is added to your table of contents and you are now ready to digitize a polygon around Carnegie Mellon University's main campus.

Trace tool

Tracing is a quick way to create new segments that follow the shapes of other features. Tracing is particularly useful when the features you want to follow have curves or complicated shapes, because snapping is more difficult in those cases.

1 On the Editor toolbar, click Start Editing > CampusStudyArea > OK.

2 Click the Create Features button.

3 Click CampusStudyArea from the Create Features panel and Polygon from the Construction panel.

4 Close the Create Features panel.

5 On the Editor toolbar, click the Trace button 🔲 ▾.

6 Click the intersection of Boundary and Forbes and drag your mouse to the right, move the cursor to the Intersection of Forbes and Morewood, follow Morewood, then Forbes again, Margaret Morrison, Tech, Frew, Schenley, and then back to Boundary.

7 When you are close to the beginning intersection of Boundary and Forbes, double-click to finish. You can click the Undo button to start over as needed. The finished polygon nicely follows straight and curved segments. Your final polygon should like the image on the right:

Cut Polygons tool

Campus planners might want the Morewood boundary to be a separate polygon. The Cut Polygons tool creates two polygons from one original polygon.

1 On the Editor Toolbar, click the Cut Polygons tool ⊞.

2 Click the left intersection of Forbes and Morewood streets, and then double-click the right intersection of Forbes and Morewood streets as shown in the image below. The result is two new polygons.

3 On the Editor toolbar click Editor > Save Edits.

Smooth tool

7-1
7-2
7-3
7-4
7-5
A7-1
A7-2

The Smooth tool smooths sharp angles in polygon outlines to improve aesthetic quality.

1 Click Bookmarks > Schenley Park.

2 On the Editor toolbar, click the Create Features button > CampusStudyArea > Polygon, and click the Straight Segment button.

3 Digitize a new polygon with 20 to 25 vertices as shown in the image.

4 On the Editor toolbar, click the Edit button and click inside the polygon.

5 On the Editor toolbar, click Editor > More Editing Tools > Advanced Editing. This adds a new Advanced Editing toolbar.

6 On the Advanced Editing toolbar, click the Smooth tool 🔘, type a maximum allowable offset of **10**, and click OK. This adds many shape vertices to create smooth curves between the polygon's vertices.

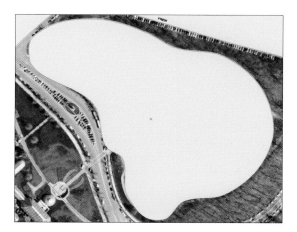

7 Click the Edit tool, double-click inside the polygon, and move vertex points to better match the park outline. **Add or delete vertex points as needed.**

8 Click outside the polygon to confirm the new shape.

Generalize tool

Generalizing creates features for use at small scales with less detail while preserving basic shapes.

1 On the Editor toolbar, click the Edit tool, and double-click inside the new traced polygon.

2 On the Advanced Editing toolbar, click the Generalize tool, type a Maximum allowable offset of **50**, and click OK. The result is a polygon with fewer vertices, no two of which have a line segment between them of less than 50 feet.

You can click the Undo button to try a different offset.

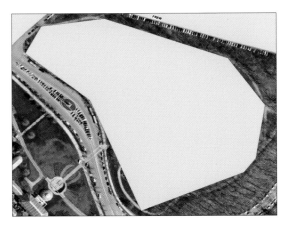

3 Click the Edit tool, double-click inside the polygon, and press DELETE.

4 On the Editor toolbar, click Editor > Stop Editing and save your edits.

Specify a segment angle and length

You have learned how to use a basemap or an aerial image as a guide to digitize new features. You can also digitize features by specifying an exact length and angle. In this exercise, you create a performance stage in the middle of Schenley Park that is exactly 80 by 40 feet.

7-1
7-2
7-3
7-4
7-5
A7-1
A7-2

1 On the Editor toolbar, click Editor > Start Editing > CampusStudy Area > OK.

2 Click CampusStudyArea in the Create Features panel and Rectangle in the Construction Tools panel.

3 Click anywhere in the middle of Schenley Park.

4 Move your cursor to start drawing a polygon, right-click, and click Direction.

5 Type **0** and press ENTER.

6 Right-click, click Length, type **80**, and press ENTER.

7 Right-click, click Width, type **40**, and press ENTER. Your rectangle is 80 feet long and 40 feet wide and its direction is to the right (ArcMap measure angles counterclockwise with zero being east or to the right).

Rotate and move a feature

The stage you created is almost finished. Next you rotate and move it to the actual location in the middle of the park.

1 Click the Edit tool and click the performance stage rectangle polygon.

2 On the Editor toolbar, click the Rotate tool [?] .

3 Click on the outline of the rectangle and drag to rotate it to match the location of the performance stage on the aerial map.

4 Click the Edit tool and move the stage on top of the stage in the aerial map.

5 Click Editor, Stop Editing, and Yes to save your edits.

6 Save your map document.

Tutorial 7-5

Spatially adjusting features

The ArcMap spatial adjustment tools transform, rubber sheet, and edge match features in a shapefile or geodatabase feature class. In this exercise, you transform an outline of a building so that it correctly overlays the building in the aerial image.

Prepare the map

1 Open Tutorial7-5.mxd from the Maps folder and save as **Tutorial7-5.mxd** to the Chapter7 folder.

2 Click the Zoom to Extent button to see both the aerial map and the building. The Hamburg Hall layer, originally a CAD drawing exported as a shapefile, is not in proper alignment or scale with the buildings shown on the aerial image. Next, you adjust the building layer so that it properly aligns with the aerial image.

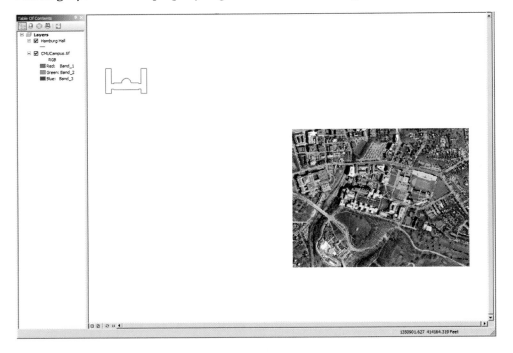

3 In the table of contents, right-click Hamburg Hall > Data > Export Data and save this as a new feature class called **HamburgHall** in Chapter7.gdb.

4 Add the new polygon feature to the map document and remove the original Hamburg Hall layer.

Move the building

1   On the Editor toolbar, click Editor, Start Editing.

Make sure the Create Features layer is Hamburg Hall.

2   Click the Edit tool, click inside the Hamburg Hall feature, and drag it to the following location on the image:

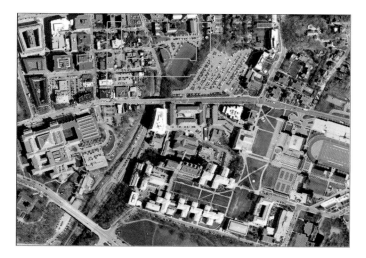

3   Zoom in to better see the Hamburg Hall feature. The building is too large and is upside down.

YOUR TURN

Rotate the Hamburg Hall feature 180 degrees.

Add displacement links

To align the feature with the aerial image, you use a transformation tool.

1   On the Editor toolbar, click Editor > More Editing Tools > Spatial Adjustment. This opens the Spatial Adjustment toolbar.

2   On the Spatial Adjustment toolbar, click Spatial Adjustment > Adjustment Methods > Transformation - Similarity.

3 Click the New Displacement Link tool .

4 Click the upper left corner of the Hamburg Hall building feature.

5 Click the corresponding location on the aerial image.

6 Continue adding displacement links to the building feature and the aerial image, as shown in the following:

Edit displacement links

If you select the wrong position on the building or map, you can use the edit displacement tools to adjust your picks.

1 From the Spatial Adjustments toolbar, click the Select Elements button.

2 Click one of the displacement links.

3 Click the Modify Link tool .

4 Drag the link to a new position.

5 Drag the link back to its original location.

YOUR TURN

Zoom to Hamburg Hall in the aerial image and use the Modify Links tool to more precisely move the displacement links to the corners of the building as shown in the graphic on the next page.

Adjust the building

1 From the Spatial Adjustment toolbar, click Spatial Adjustment > Adjust.

2 Stop editing and save your edits. ArcMap scales down the Hamburg Hall feature to match the geometry of the feature in the aerial image. If the resulting match is not as good as you would like, select the Hamburg Hall feature, redefine new displacement links, and run the Adjust command again.

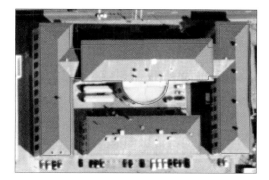

3 Save your map document and exit ArcMap.

Assignment 7-1

Digitize police beats

Community-oriented police officers are responsible for preventing crime and solving underlying community problems related to crime. Among other activities, these officers walk "beats," which are small networks of streets in specified areas. Often the beats are designed in cooperation with community leaders who help set policing priorities. Beats change as problems are solved and priorities change. Hence, it is good to have the capability to digitize and modify police beats.

In this assignment, you digitize two new polyline police beats for the City of Pittsburgh Zone 2 Police District based on street centerlines that make up these beats.

Get set up

- Rename the folder \EsriPress\GIST1\MyAssignments\Chapter7\Assignment7-1YourName\ to your name or student ID. Store all files that you produce for this assignment in this folder.
- Create a new map document called **Assignment7-1YourName.mxd** with relative paths.
- Create a new file geodatabase called **Assignment7-1YourName.gdb**.

Build the map

Import or copy the following data into your file geodatabase:

- \EsriPress\GIST1\Data\Pittsburgh\Zone2.gdb\Streets—line features for Zone 2 Police District streets.
- \EsriPress\GIST1\Data\Pittsburgh\Zone2.gdb\Outline—polygon feature for an outline boundary of the Zone 2 Police District.

Create and digitize new features

In your file geodatabase create a new polyline feature class called **PoliceBeats** and assign it the same coordinate system as the Zone2 features, NAD_1983_StatePlane_Pennsylvania_South_FIPS_3702_Feet. Create a new text field called **BeatNumber**. Digitize line segments for the police beats using the guidelines that follow for what streets make up the beats. Populate the fields as Beat1 and Beat2 for each line segment making up that beat.

Hints: Open the feature attribute table for the streets. Move the table so you can see both the table and the streets on the map. Sort the table by field, NAME, and make multiple selections for a given beat in the table by simultaneously holding down the CTRL key and clicking rows corresponding to the beat's street segments. The streets layer is a TIGER file map with TIGER-style address number data, so look for street number ranges in the following fields: L_F_ADD, L_T_ADD, R_F_ADD, and R_T_ADD. With all streets for a beat selected, digitize lines for streets making up beats in the new line features for Beat1 and Beat2. Use the snapping endpoint and vertex tools using street centerlines as a guide.

Street centerline guides for Beat #1 (21 streets)

1 through 199—17th St (four segments)

1 through 99—18th St (two segments)

1 through 199—19th St (one segment)

1 through 199—20th St (four segments)

1 through 99—Colville St (one segment)

1700 through 1999—Liberty Ave (one segment)

1700 through 1999—Penn Ave (three segments)

1700 through 1999—Smallman St (four segments)

1700 through 1999—Spring Way (one segment)

Street centerline guides for Beat #2 (20 streets)

100 through 299—7th St (two segments)

1 through 299—8th St (three segments)

100 through 299—9th St (four segments)

800 through 899—Exchange Way (one segment)

700 through 899—Ft Duquesne Blvd (three segments)

700 through 899—Liberty Ave (three segments)

100 through 199—Maddock Pl (one segment)

700 through 899—Penn Ave (three segments)

Requirements

- Create a layout with three data frames, renaming the data frames and layers. Using Unique Values, show the beats with line widths 1.5 and bright, distinctive colors (Mars Red and Ultra Blue); and show streets as lighter "ground" features.

- In data frame 1, show an overview map zoomed to the Police Zone 2 outline, with existing streets and the newly digitized beats labeled as Beat1 and Beat2 using a callout label from the Draw toolbar.

- In data frames 2 and 3, include maps zoomed to Beat 1 (dataframe 2) and Beat2 (dataframe 3) at the same scale. Label the streets in the detailed maps.

- Include scale bars in miles for the overview map and in feet for the individual beats' maps. Include your name as the author and the current date.

- Click File > Export Map and save your map layout as **Assignment7-1YourName.jpg** with 150 dpi resolution.

Assignment 7-2

Use GIS to track campus information

GIS is a good tool to create campus information or emergency preparedness maps. These maps can be used in many organizations that have large campuses or complicated buildings. For example, airports, hospitals, office parks, or colleges and universities can be confusing, especially to new students and visitors. Visitors might also be unaware of locations of features that would be useful for emergency purposes.

In this assignment, you create spatially modified and new features for sidewalks, parking lots, handicap parking, and emergency management service locations. These features are shown in the image for this assignment, and you can visit Carnegie Mellon University's visitor map for additional parking and emergency service information: http://www.cmu.edu/about/visit/campus-map.shtml.

Get set up

- Rename the folder \EsriPress\GIST1\MyAssignments\Chapter7\Assignment7-2YourName\ to your name or student ID. Store all files that you produce for this assignment in this folder.
- Create a new map document called **Assignment7-2YourName.mxd** with relative paths.
- Create a new file geodatabase called **Assignment7-2YourName.gdb**.

Build the map

Import or copy the following data and raster image into your file geodatabase:

- \EsriPress\GIST1\Data\Pittsburgh\City.gdb\PghStreets—street centerline features for Pittsburgh.
- \EsriPress\GIST1\Data\CMUCampus\Walkways.dwg—CAD drawing of campus sidewalks (originally extracted from CMU's Facilities Management Services campus drawing). Use the line features of this drawing file.
- \EsriPress\GIST1\Data\CMUCampus\CMUCampus.tif—USGS ortho image of Carnegie Mellon University's campus.

Create new features

In your file geodatabase, create the following feature classes using the same map projection as PghStreets features, NAD_1983_StatePlane_Pennsylvania_South_FIPS_3702_Feet:

- **ParkingLots** (polygons) with a new text field called **LOT_NAME**
- **HandicapParking** (points)
- **EMS_Sites** (points) with a new text field called **NAME**

Requirements

- Use the Move and Rotate Vertex tools to display the CMU sidewalks to the approximate locations shown as white lines in the next figure's aerial image.
- Digitize the five parking lots shown in the image and enter the corresponding parking lot names. Label the lot name on your map. Display the parking lots as semitransparent polygons so you can see the parked cars in the aerial image.
- Digitize eleven handicap parking locations as shown in the image using the Esri "Handicapped2" symbol.
- Digitize EMS locations for Health Services and University Police as shown and enter corresponding names in the attribute table. Display the points as unique symbols using the Esri "Cross 4" and "Asterisk 2" symbols.
- Display PghStreets with no color but labeled using a small halo mask.

7-1
7-2
7-3
7-4
7-5
A7-1
A7-2

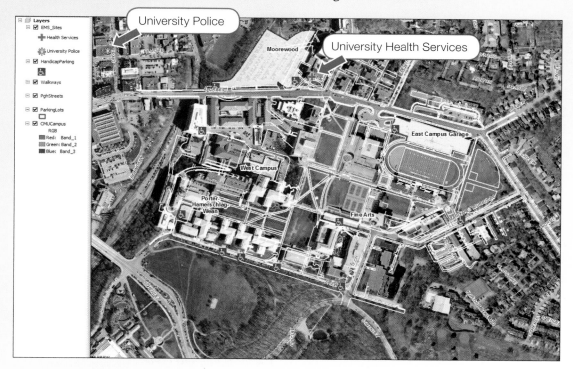

- Create a map layout zoomed to the above features at a scale of 1:4,000. Include a title, legend, map scale using feet, your name, and the current date.
- Click File > Export Map and save your map layout as **Assignment7-2YourName.jpg** with 150 dpi resolution.

Part II
Working with spatial data

8

Geocoding

Geocoding is the process used to plot address data as points on a map. You can geocode addresses to different levels such as ZIP Codes or streets, depending on the type of address data you have or wish to map. In this chapter, you learn to geocode using source tables of address data and location reference data obtained from the US Census Bureau. You also learn how to fix errors in both the source and reference data you use for geocoding.

Learning objectives

- Geocode by ZIP Code
- Geocode by street address
- Correct address data interactively
- Edit street map layers to improve geocoding
- Use an alias table for place-names

Tutorial 8-1

Geocoding data by ZIP Code

Geocoding to ZIP Codes is a common practice for many organizations because ZIP Code data is often available in client and other databases. Furthermore, for marketing and planning it is often sufficient to study the spatial distribution of clients by ZIP Code. ZIP Code areas lack an underlying design principle except for delivery of mail, so interpretation of results is sometimes limited. In these tutorials, you geocode attendees for an art event sponsored by an arts organization in Pittsburgh, Pennsylvania, called FLUX. The event planners of FLUX would like to know where function attendees reside for planning and marketing activities.

Open and examine the starting map document

1 Start ArcMap and open Tutorial8-1.mxd from the Maps folder. The map document includes the two needed inputs for geocoding, the ZIP Code map of Pennsylvania, and the data table Attendees.

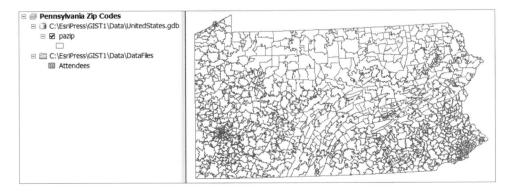

2 Save the map document to the Chapter8 folder of MyExercises.

3 In the table of contents, right-click the Attendees table and click Open. The table contains the addresses, including 5-digit ZIP Code, and ages of all attendees of two recent FLUX events. Notice that two of the first six attendees are out of state and thus will not geocode with the Pennsylvania ZIP Code map. As a supplement, before closing the table, let's tabulate how many such records there are. That informs the performance assessment of geocoding within Pennsylvania.

Attendees							
OID	Date_	Address	City	State	ZIP_Code	Age	
0	20090629	2415 1ST AVE	SACRAMENTO	CA	95818	27	
1	20090629	224 NORTH ST	STERBENVILLE	OH	43952	32	
2	20090629	PO BOX 622 535 4TH ST	MARIANNA	PA	15345	32	
3	20090629	5126 JANIE DRIVE	PITTSBURGH	PA	15227	34	
4	20090629	305 AVENUE A	PITTSBURGH	PA	15221	40	
5	20090629	1431 CRESSON ST	PITTSBURGH	PA	15221	26	

4 Using the Table Options button, click the Select By Attributes button, double-click "State" in the top panel > the = button, click Get Unique Values, scroll down and double-click 'PA', and click Apply > Close. Information at the bottom of the table indicates that 1,124 out of 1,265 records (89%) are for Pennsylvania, so that is the maximum number of points that could be geocoded. If there are missing or incorrect ZIP Codes for Pennsylvania, the number geocoded will be less.

5 Using the Table Options button, click the Clear Selection button and close the table.

Create an address locator for ZIP Codes

The geocoding process requires several settings and parameters. Rather than have you specify them interactively each time you geocode data, ArcMap has you save settings in a reusable file, called an address locator. Included in the settings is a pointer to the reference data you use to geocode the attendee data, the PAZip (Pennsylvania ZIP Code) layer that is currently in your map document.

1 In the Catalog window, navigate to the Chapter8 folder of MyExercises, right-click Chapter8.gdb, and click New and Address Locator.

2 In the Create Address Locator window, click the browse button for the Address Locator Style, click US Address–ZIP 5-Digit, click OK, and ignore the warning icon and message.

3 Click the list arrow for Reference Data, and click pazip.

4 For the Output Address Locator, change the file name from PAZip_CreateAddressLocator to **PAZipCodes** for name.

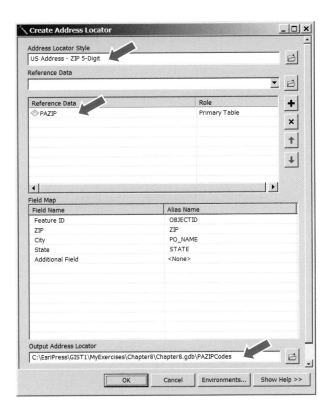

5 Click OK and wait until Catalog informs you that the address locator is created. PAZipCodes appears as an address locator under Chapter8.gdb in Catalog.

6 Hide the Catalog window.

Geocode records by ZIP Code

1 In ArcMap, click Customize > Toolbars > Geocoding. The Geocoding toolbar appears.

2 On the Geocoding toolbar, click the Geocode Addresses button 🐾 , click PAZipCodes to select it > OK. Check that the Attendees table is selected as the Address table, select ZIP_Code as the ZIPCode field, and change the name of the output to **\EsriPress\ GIST1\MyExercises\Chapter8\Chapter8.gdb\AttendeesZIP**.

3 Click OK. ArcMap geocodes the addresses by ZIP Code with 86 percent of the records mapped—less than the 89 percent of addresses that are in Pennsylvania, as expected.

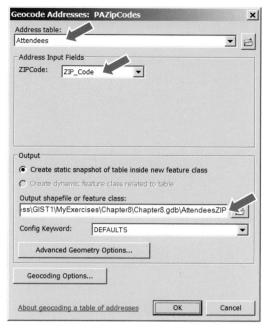

8-1
8-2
8-3
8-4
8-5
A8-1
A8-2
A8-3

4 Click Close. ArcMap adds the geocoding results to the map with point markers at the centroids of ZIP Codes that have one or more attendees. As you might expect, attendees cluster around southwestern Pennsylvania near the location of the FLUX events, but attendees come from all over the state.

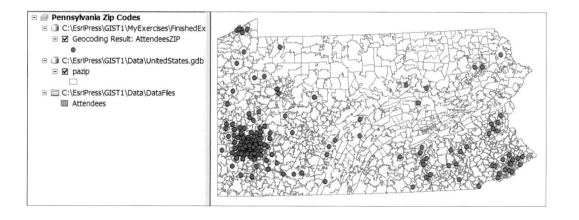

Count geocoded records by ZIP Code

You can get a better understanding of the geocoded output by next taking an extra step, aggregating geocoded points to obtain a count of attendees per ZIP Code area.

1 In the table of contents, right-click GeocodingResult: AttendeesZIP and click Open Attribute Table.

2 Right-click the Match_Addr column header, click Summarize, change the output table to **\EsriPress\GIST1\MyExercises\Chapter8\Chapter8.gdb\CountAttendees**, and click OK > Yes.

3 Close the geocoding results table.

4 Right-click CountAttendees in the table of contents and click Open.

5 Right-click Count_Match_Addr column header and click Sort Descending. In total there are 1,487 ZIP Codes in the state (which you can determine by opening the attribute table of pazip), but only 22 ZIP Codes with 10 or more attendees (1.5%) and 90 (6.1%) that have 2 or more attendees. This is valuable information for FLUX for targeting marketing activities.

6 Close the table.

YOUR TURN

Join CountAttendees to GeocodingResult:AttendeesZIP using Join attributes to a table. In the Join Data window, you have to use the option to join attributes from another table, select ZIPCode for item 1, CountAttendees for item 2, and match_addr for item 3. Create a map displaying Count_Match_Addr using size-graduated point markers (graduated symbols in the show panel of the Symbology tab) with the Count_Match-addr field, five classes and quantiles. Zoom in to southwestern Pennsylvania.

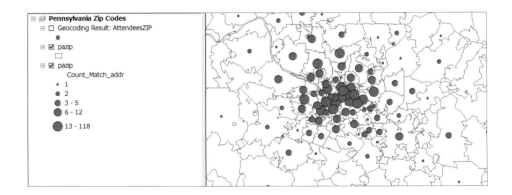

Fix and rematch ZIP Codes

8-1
8-2
8-3
8-4
8-5
A8-1
A8-2
A8-3

1 In the table of contents, click Geocoding Result: AttendeesZIP and click the Review/ Rematch Addresses button 🖳 on the Geocoding toolbar. The Interactive Rematch window shows each unmatched record individually and allows you to manually edit the address values.

2 For Show results select Unmatched Addresses, scroll down to the record with ObjectID=50, whose address value is 414 South Craig Street, and select that record.

In the next step, attribute names are fully qualified, meaning that table name is added as a prefix to attribute name using dot notation. That is because the previous Your Turn assignment had you join CountAttendees to GeocodingResult:AttendeesZIP and ArcGIS is including the table name for each attribute to clarify source. If you did not do the Your Turn assignment, you will just have attribute names without prefix table names.

3 Scroll horizontally (and adjust field widths by dragging their header boundaries) so you can see the AttendeesZIP.Address, AttendeesZIP.City, AttendeesZiP.State, and AttendeesZIP.ZIP_Code fields. Notice that the AttendeesZIP.ZIP_Code is missing for this record. That address's ZIP Code is 15213.

4 With the 414 SOUTH CRAIG STREET row selected, type **15213** in the ZIPCode field (top of bottom, left panel) and press TAB on your keyboard. The Candidate panel shows one candidate with a perfect score of 100.

5 Click the Match button. The count of matched addresses goes up by one, from 1,090 to 1,091, as ArcMap is successful with that point. With some research, it is usually possible to make similar additions.

YOUR TURN

Use the US Postal Service's ZIP Code lookup website, http://zip4.usps.com/zip4/welcome.jsp, to find the ZIP Code for the record with ObjectID=57, 11244 Azalea Dr, Pittsburgh, PA (it's 15235). Then use Interactive Rematch to match the address. Close the Interactive Rematch window, and save your map document when finished.

Tutorial 8-2

Geocoding data by street address

In this exercise, you again geocode the FLUX attendee records, but this time at the street level for Pittsburgh. In this case, you incorporate ZIP Code into the address locator, because some addresses may have the same house number and street name but be in different ZIP Code areas. This happens frequently in study areas that have two or more municipalities, such as in a county. In the FLUX case there are address records from other cities and states, so ZIP Code plays an important role, given the preponderance of Pittsburgh records, to eliminate matches with non-Pittsburgh street addresses that have the same street address as in Pittsburgh.

Examine address data and street map

1 Open Tutorial8-2.mxd from the Maps folder and save the map document in the MyExercises Chapter8 folder.

2 Click the List By Source button at the top of the table of contents. Open the attribute tables for PghStreets and Attendees and review their contents, especially addresses. You will find that PghStreets has TIGER-style street address data, with starting and ending house numbers for each street segment. The table, Attendees, has street address in one field, Address, plus City, State, and ZIP_Code in their own fields. Only Address and ZIP_ Code are necessary for geocoding, because with ZIP_Code you can look up city and state.

3 Close the tables after you are finished reviewing them.

Create an address locator for streets with a zone

1 Click the Catalog icon in the right edge of the ArcMap window (or, if the icon is not there, click Windows > Catalog).

2 Expand the Chapter8 folder of MyExercises, right-click Chapter8.gdb, and click New > Address Locator.

3 In the Create Address Locator window, click the browse button for the AddressLocator Style, click US Address–Dual Ranges, click OK, and ignore the warning icon and message.

4 Select PghStreets for Reference Data, and change the Output Address Locator name to **PghStreets**. Notice that Catalog identifies all fields that it needs for the geocoding process in the lower panel of the Create Address Locator window. If it were unsuccessful in doing so, you would have to click in a cell on the right and select the needed field name from the table. Field names that start with an asterisk are required.

5 Click OK and wait until Catalog informs you that the address locator is created.

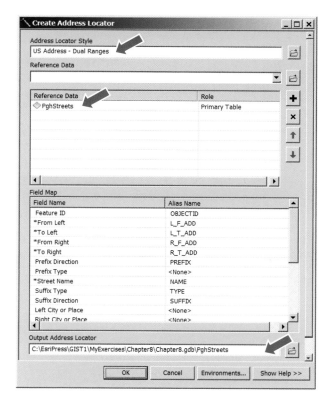

6 Expand Chapter8.gdb in the Catalog tree and double-click PghStreets. Expand all items in the resulting property sheet. See image on next page. In the Place name alias table item, you can associate an alias table with the locator. It would contain place-names such as PNC Ballpark and associated street addresses such as 115 Federal Street. With an alias table, ArcMap makes a pass through address data, replacing place-names with their street addresses. In Outputs, you can have x and y coordinates written to the output attribute table. In Geocoding options are parameters with default values that determine how strict or lax addresses must compare in the source and reference datasets to be considered a match. Work Assignment 8-3 to learn how to modify these parameters. Also in this element, you see the allowable connectors for street intersection addresses, such as Oak St & Pine Ave, are currently the "&", "@", and "|" characters, and the words "and" and "at". If your data has a different separator, you could type it here. Also, the point that ArcMap assigns to an address has a 20-foot side offset on the correct side of the street (left or right). You can change the offset to another value if desired. Offsets are desirable, especially for aggregating geocoded data points up to counts by area, such as neighborhoods. Areas tend to use street centerlines as boundary lines, so an offset ensures that ArcMap will count each point once, in the correct polygon. If on the centerline, a point gets double-counted for polygons that share the street segment as a boundary line.

7 Click Cancel to close the Address Locator Properties window, and hide the Catalog window.

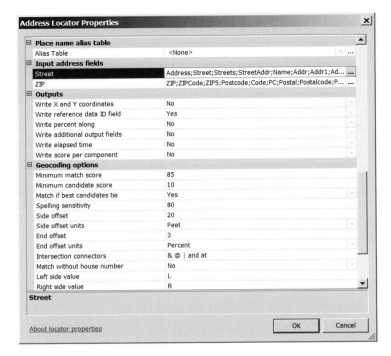

Interactively locate addresses

Before geocoding the Attendees table to the streets layer, you try out your locator with ArcMap's Find tool to locate individual addresses. The Find tool that you use has the same methodology as geocoding to transform street address data into a point on the map. It matches the address you type in with similar data stored as attribute data in the street centerline map. It does the matching by finding good candidates and then computing a match score for each. Then for each identified problem or flaw of a candidate in matching the desired address, ArcMap subtracts a penalty from the match score. The candidate with the perfect score, 100, or highest score above a threshold value is chosen as the geocoded point.

1 On the Tools toolbar, click the Find button .

2 Click the Locations tab. ArcGIS should automatically find your PghStreets locator. If not, click the browse button for locator, and browse to and select your PghStreets locator in Chapter8.gdb.

3 Type **3609 Penn Ave 15201** in the Full Address field and click Find. The locator finds the address, briefly flashing it on the map. The address locator works well.

4 Right-click the matched address of the lower panel to open a context menu and click the Add Labeled Point option. A point appears at the corresponding point on the map, along with a label for the street address.

5 Close the Find window.

◎ Flash
⊕ Zoom To
⟨ᵐ⟩ **Pan To**
▣ Create Bookmark
⊙ Add Point
⊟ Add Labeled Point
▭ Add Callout
▷ Add to My Places
▷ Manage My Places...
⟨• Add as Stop to Find Route
⟨• Add as Barrier to Find Route
Add as Network Analysis Object
Move Network Analysis Object

YOUR TURN

Use the Find tool to locate the following addresses. Add the best match in each case to the map as a labeled point.

- 1920 S 18th ST 15203
- 255 Atwood ST 15213
- 3527 Beechwood BLVD 15217

The finished map below has some editing of the labels, which is optional for you. You can ungroup a label and its point by clicking the label to select the graphic, right-clicking the graphic, and clicking Ungroup. Then you can format the font and point color by right-clicking an element, clicking Properties, and clicking Change Symbol.

When finished, delete the label and point graphics by dragging a rectangle around them to select them and press DELETE on your keyboard.

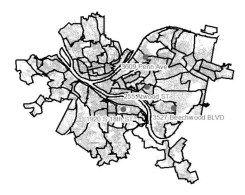

Geocode address data to streets

We do not expect ArcMap to geocode a high percentage of matched records because the streets reference layer is only for Pittsburgh while many attendees live outside of Pittsburgh. Also, note that your matched addresses and statistics may not match those shown here because the Esri software developers frequently improve ArcGIS locators, even with releases of software service packs.

1 If the Geocoding toolbar is not already open, click Customize > Toolbars > Geocoding.

2 On the Geocoding toolbar, click the Geocode Addresses button, select PghStreets, and click OK.

3 Type or make selections as shown in the image (but do not click OK).

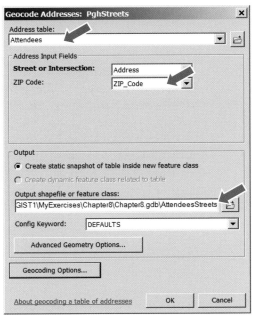

4 Click Geocoding Options. This is a user dialog box controlling the behavior of address matching and its outputs. You saw several of these options earlier. Here you can choose to add the X and Y coordinates of matched points to the output map layer's attribute table; however, leave all default settings as found here.

5 Click Cancel, OK. ArcMap only matches 550 (43%) of records, a relatively low number as expected, because there are many addresses outside of Pittsburgh as well as unmatched addresses in Pittsburgh. You can count 794 Pittsburgh records in the Attendees table, so the match rate for Pittsburgh is low, 550 out of 794 or 70%. Sometimes street address data collected in surveys is low quality as seems to be in the case here.

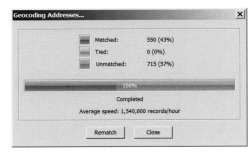

6 Click Close. The residences of attendees have distinctive spatial patterns across Pittsburgh neighborhoods, which could inform marketing campaigns for increasing attendance. It looks like most attendees come from only 15 to 20 out of the 91 neighborhoods comprising Pittsburgh.

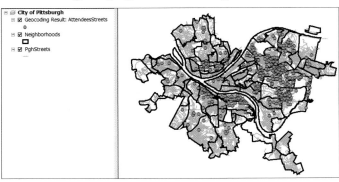

8-1
8-2
8-3
8-4
8-5
A8-1
A8-2
A8-3

Tutorial 8-3

Correcting source addresses using interactive rematch

While ArcMap did not match many of the addresses in the Attendees table to the PghStreets layer because they are outside of Pittsburgh, some addresses in Pittsburgh did not match due to spelling errors or data omissions in the input table. Making corrections to the address data depends on the user's knowledge of local streets and addresses. In this tutorial, you use the ArcMap interactive review process to correct and then match a few of the unmatched records.

Rematch interactively by correcting input addresses

1 Save your map document as **Tutorial8-3.mxd** to the Chapter8 folder.

2 Click Geocoding Result: AttendeesStreets in the table of contents. On the Geocoding toolbar, click the Review/Rematch Addresses button. You can increase the height of the panel with addresses by finding and dragging horizontal boundary lines.

3 In the Show results field, select Unmatched Addresses.

4 Horizontally scroll across the fields so you can see the Address, City, State, and Zip_ Code fields, and then make those fields narrower by dragging vertical boundaries between fields so that you can see all of their values.

5 Right-click the Address header in the data panel and click Sort Ascending.

6 Scroll down to the record with address 120 W 9th STREET and select that record. ArcMap did not match this address because its direction is wrong and its ZIP Code is wrong. The correct direction is S, not W and the ZIP Code should be 15203.

7 Change the direction from W to **S** in the Street or Intersection field and type **15203** in the ZIP Code field of the lower left panel, press the TAB key, and click Match. ArcMap matches the records and the count of Matched addresses advances by one from 550 to 551.

> ### *YOUR TURN*
>
> Rematch an additional record, 1715 CENTER AVE. The problem here is that CENTER is spelled wrong, it should be CENTRE, and the ZIP Code is wrong, it should be 15219. Make the corrections in the lower left panel of Interactive Rematch and click Match. The number matched will advance to 552. Leave the Interactive window open.

Rematch interactively by pointing on the map

Sometimes you will have unmatched records that you can find on the map using external information or expert knowledge, but your reference data (street map) simply will not have a corresponding address or street. Often too, TIGER street maps have missing house numbers or other data for street segments so that address matching is not possible. In such cases, ArcMap lets you point on the map to geocode.

1 If the Interactive Rematch window is not open, click Geocoding Result: AttendeesStreets in the table of contents. Then on the Geocoding toolbar, click the Review/Rematch Addresses button. Select Unmatched Addresses for the Show results field.

2 Sort the address data by ObjectID, scroll down to the record with ObjectID 1017, and click that record to select it. While this record has only a ZIP Code value, 15221, for address data, suppose that a comment field of the original survey data mentioned that the attendee lived on Canada Way. That street is only one block long in Pittsburgh, so you decide to point to the middle of that street to address match the point.

3 On the Tools toolbar, click the Zoom In tool and zoom in to the eastern portion of Pittsburgh as seen in the image. Highlighted is Canada Way. If you have difficulty locating that street, you can use any street segment in its vicinity for practice here.

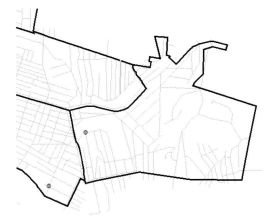

4 On the bottom of the Interactive Rematch window, click Pick Address from Map, move your cursor to the middle of Canada Way, right-click, click Pick Address, and scroll to the left in the address panel of the Interactive Match window. ArcMap adds the address as a point on the map where you clicked. It sets the record's Status to M for matched and sets the Match_type to PP for picked point (for other records, A is the code for address matched).

5 Close the Interactive Rematch window and click the Full Extent button on the Tools toolbar.

8-1
8-2
8-3
8-4
8-5
A8-1
A8-2
A8-3

Tutorial 8-4

Correcting street reference layer addresses

In this tutorial, you learn how to find and fix an incorrect address in a reference street layer used for geocoding. To do this, you examine unmatched user addresses, identify candidate streets for revisions, and examine the attributes of the streets to look for misspellings or data omissions.

Open a map document

1 Open Tutorial8-4.mxd from the Maps folder. Tutorial8-4 contains a table of clients and a street centerlines layer for Pittsburgh's Central Business District.

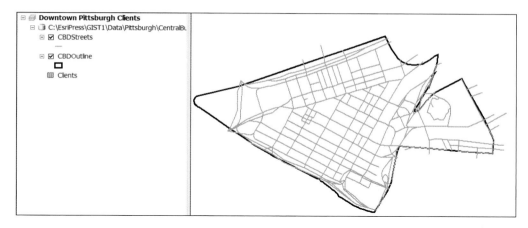

2 Save the map document to the Chapter8 folder.

3 Right-click Clients in the table of contents and click Open. The data table has addresses for all 27 records, including ZIP Code. Notice that the last four records have place-names instead of street addresses. Later in the tutorial you use an alias table to geocode those locations. The alias table includes street addresses for place-names.

	OBJECTID *	LAST_NAME	FIRST_NAME	ADDRESS	ZIP
▶	1	Roberts	Louisa	133 Seventh St	15222
	2	Johnson	Tammy	615 Penn Ave	15222
	3	Clark	Robert	309 Ross St	15222
	4	Peterson	Jennifer	118 6th Street	15222
	5	Young	Mike	490 Penn Ave	15222
	6	Thompson	Samantha	711 Liberty Ave	15222
	7	Reed	Rhonda	111 Hawksworth	15222
	8	Baker	Sally	900 Smallman St	15222
	9	Wilson	Laura	599 Smithfield St	15222
	10	Jenkins	Amy	701 Grant Street	15222
	11	Kelly	Tabitha	900 Lib Ave	15222
	12	Riley	Jennifer	777 Illini Drive	15222
	13	Smith	Emily	341 Stanwix St	15222
	14	Williams	Polly	109 Washington Pl	15222
	15	Davis	Kelly	1100 Liberty Ave	15222
	16	Dobbins	Joshua	2 S. Market Place	15222
	17	Perry	Catherine	651 Forbes Ave	15222
	18	Nelson	Joseph	241 Forbes Ave	15222
	19	Miller	Matthew	923 French St	15222
	20	Lawson	Todd	295 Wood St	15222
	21	Welch	Karen	301 5th Ave	15222
	22	Sigler	Dan	119 9th Avenue	15222
	23	Hampton	Frances	401 1st Ave	15222
	24	Peters	Amanda	One PPG Place	15222
	25	Franklin	John	One PPG Place	15222
	26	Burns	Anthony	Two Gateway Center	15222
	27	Smith	George	Two PPG Place	15222

4 Close the table.

Create an address locator for CBD streets

1 Click the Catalog icon in the right edge of the ArcMap window (or if the icon is not there, click Windows > Catalog).

2 Expand the Chapter8 folder of MyExercises, right-click Chapter8.gdb, and click New and Address Locator.

3 In the Create Address Locator window, click the browse button for the Address Locator Style > US Address–Dual Ranges > OK, and ignore the warning icon and message.

4 Select CBDStreets for the Reference Data.

5 Click the browse button for Output Address Locator, and change the locator name to **PghCBDStreets** in the Output Address Locator field.

6 Click OK.

7 Hide the Catalog window.

Geocode clients' addresses to CBD streets

1 If the Geocoding toolbar is not open, click Customize > Toolbars > Geocoding.

2 On the Geocoding toolbar, click the Geocode Addresses button.

3 Select PghCBDStreets and click OK.

4 In the Geocode Addresses window, type **\EsriPress\GIST1\MyExercises\Chapter8\ Chapter8.gdb\CBDClients** for Output shapefile or feature class, and click OK. ArcMap matches 12 (44 percent) of the 27 records.

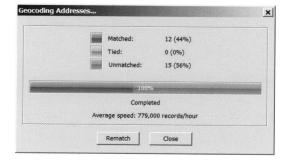

5 Click Close.

Identify a problem street segment record using Review/Rematch Addresses

1 Click Geocoding Result: CBD Clients in the table of contents to select it.

2 On the Geocoding toolbar, click the Review/Rematch Addresses button. Select Unmatched Addresses for the Show results field.

3 Scroll to the right in the unmatched addresses, right-click the ADDRESS column header, and click Sort Ascending.

4 Scroll down and select the record with ADDRESS 490 Penn Ave. Click the Geocoding Options button, type 0 for the Minimum candidate score, and click OK. There are many candidate street matches in the lower panel of the Interactive Rematch window. The closest match is 500 PENN AVE 15222.

5 Select 500 PENN AVE 15222 in the Candidates panel and zoom in to the area with the yellow point marker on the map. Street address numbers increase from left to right in the CBD, and you can see in the lower right of the Interactive Rematch window that the best candidate's lowest number is 500. So the desired street segment must be to the immediate left of the yellow point marker.

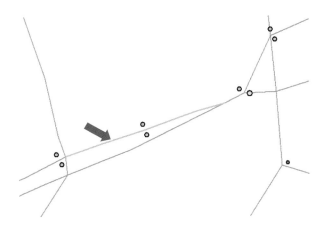

6 On the Tools toolbar, click the Select button and click the street segment indicated in the graphic above.

7 Close the Interactive Rematch window.

8 Open the CBDStreets attribute table, and click the Selected button at the bottom of the table.

9 Scroll to the right in the table to see that the selected street segment's record is missing its TIGER-style street numbers (from and to, left and right street numbers). Leave the table open.

RECNUM	L_F_ADD	L_T_ADD	R_F_ADD	R_T_ADD	FCC
24924586					A35

Edit a street record

Suppose that you have obtained valid numbers for the street segment's missing attributes: 498 to 474 on the left side and 499 to 475 on the right side. You can use the ArcMap Editor toolbar to enter those values.

1 On the Menu bar, click Customize > Toolbars > Editor.

2 Click the list arrow on the Editor toolbar. Click Start Editing > CBDStreets > OK.

3 In the CBDStreets table, type the following values: L_F_ADD=**498**, L_T_ADD=**474**, R_F_ADD=**499**, and R_T_ADD=**475**.

4 Click the list arrow on the Editor, click Save Edits > Stop Editing.

5 Close the CBDStreets table and the Editor toolbar. Now, 490 Penn Ave., one of the unmatched client addresses, will geocode the next time you attempt to rematch the addresses.

First, however, you have to rebuild the PghCBDStreets Locator so that it includes the edits you just made to the reference CBDStreets layer.

Rebuild a street locator

If you modify a location reference layer after you built a locator, you have to rebuild the locator.

1 Unhide or open Catalog.

2 If necessary, expand the Chapter8 folder of MyExercises and Chapter8.gdb.

3 Right-click PghCBDStreets and click Rebuild, OK.

4 Hide Catalog.

Rematch interactively using an edited street segment

1 Click Geocoding Result: CBD Clients in the table of contents to select it.

2 On the Geocoding toolbar, click the Review/Rematch Addresses button, select Unmatched Addresses for the Show results field, and select the 490 Penn Ave. address. Now there is a 100 score candidate for geocoding, for exactly the right address, 490 PENN AVE 15222, given your updated street segment.

3 In the Interactive Rematch window, click the 100 score candidate record > Match. ArcMap matches the record, and the number of matched records increases from 12 to 13. In addition, any time you use CBDStreets in the future for geocoding, ArcMap will successfully geocode any input address records for the edited street segment.

4 Close the Interactive Rematch window.

5 Save your map document.

Tutorial 8-5

Using an alias table

Some places are commonly located by their landmark names instead of their street addresses. For example, the White House may be listed in a table as "White House" instead of 1600 Pennsylvania Avenue NW, Washington, D.C., 20500. In the following exercise, you use an alias table to geocode records that are identified by their landmark name rather than their street address.

Add an alias table and rematch addresses

1 Save your map document as **Tutorial8-5.mxd**.

2 Click the Add Data button and add the BldgNameAlias table from Data > Pittsburgh > CBD.gdb.

3 In the table of contents, right-click BldgNameAlias and click Open. The table contains the alias names and street addresses for three records. You can create such tables in Microsoft Excel, in Notepad with comma delimiters, or in other packages, and import them into a file geodatabase using Catalog.

OBJECTID *	BLDGNAME	ADDRESS	ZIP
1	One PPG Place	200 4th Ave	15222
2	Two PPG Place	200 4th Ave	15222
3	Two Gateway Center	197 Stanwix St	15222

4 Close the alias table.

5 Click Geocoding Result: CBD Clients in the table of contents to select it. Click the Review/Rematch Addresses button on the Geocoding toolbar. Select Unmatched Addresses for the Show results field.

6 Click Geocoding Options (bottom left of window) > Place Name Alias Table.

7 Select BldgNameAlias for the Alias table, BLDGNAME for the Alias field, and click OK > OK.

8 Scroll down in the address records, click the record with ObjectID 24, and scroll to the right to see the address—One PPG Place. Now that you have an alias for that building, there is a candidate with a 100 score.

9   Click the candidate with score 100 and click Match. Similarly, match the last three records.

10   Close the Interactive Rematch window and close ArcMap.

8-1
8-2
8-3
8-4
8-5
A8-1
A8-2
A8-3

Assignment 8-1

Geocode household hazardous waste participants to ZIP Codes

Many county, city, and local environmental organizations receive inquiries from residents asking how they can dispose of household hazardous waste (HHW) materials that cannot be placed in regular trash or recycling collections. Homeowners continually search for environmentally responsible methods for disposing of common household products such as paint, solvents, automotive fluids, pesticides, insecticides, and cleaning chemicals.

The Pennsylvania Resources Council (PRC) (http://www.prc.org/) is a nonprofit organization dedicated to protecting the environment. The PRC facilitates meetings, organizes collection events, spearheads fundraising and volunteer efforts, and develops education and outreach materials in response to the HHW problem.

At each disposal event, the PRC collects residence data from participants. In this assignment, you geocode participants by ZIP Code for a recent Allegheny County event with collection of hazardous waste that took place in the southeastern portion of Allegheny County's North Park. The beauty of ZIP Code data is that everyone knows their five-digit ZIP Code so that you can get highly accurate and nearly complete data that geocodes well.

Get set up

- Rename the folder \EsriPress\GIST1\MyAssignments\Chapter8\Assignment8-1YourName\ to your name or student ID. Store all files that you produce for this assignment in this folder.
- Create a new map document called **Assignment8-1YourName.mxd** with relative paths. Use the UTM NAD1983 Zone 17N projection.
- Create a new file geodatabase called **Assignment8-1YourName.gdb**. Save all new files that you create in this file geodatabase.

Build the map

Add the following spatial data to your map document:

- \EsriPress\GIST1\Data\UnitedStates.gdb\HHWZIPCodes—table of five-digit ZIP Codes for a HHW Allegheny County event provided by the PRC. Note that the PRC suppressed all attributes except ZIP Code to protect confidentiality.
- \EsriPress\GIST1\Data\UnitedStates.gdb\PAZip—polygon features of Pennsylvania ZIP Codes used for address matching.
- \EsriPress\GIST1\Data\UnitedStates.gdb\PACounties—polygon features of Pennsylvania counties.
- \EsriPress\GIST1\Data\AlleghenyCounty.gdb\Parks—polygon features of Allegheny County parks.

Geocode HHW ZIP Codes:

- Create an address locator to use for geocoding HHW participants to ZIP Codes, **HHWZIPLocator**.

- Geocode a new map layer, HHWZIPCodeResidences.

- Spatially join HHWZIPCodeResidences to PAZIP. *Hint:* Right-click PAZIP, click Joins And Relates > Join and select join data from another layer based on spatial location.

Requirements

- Symbolize your map including a choropleth map in a layout showing the number of Household Hazardous Waste participants by ZIP Code in Pennsylvania. Use five quantiles for the choropleth map. Add the PA County polygons as a thick dark outline. Label counties with county names. Add a second copy of PAZip with outline and white fill to complete ZIP Codes throughout Pennsylvania where there were no HHW client residences. *Hint:* Select North Park and place Parks below the choropleth map, to see only North Park's location. Add text to your layout to label North Park.

- Export a layout with title, map, and legend to a file called **Assignment8-1YourName.jpg**. Include the layout image in a Word document, **Assignment8-1YourName.docx**, in which you describe residence patterns of HHW event attendees relative to North Park. Include a statement about the top 20 percent of clients. Roughly speaking, how far were they willing to travel; 5, 10, 15, or 20 miles? Include geocoding match statistics in your Word document.

Assignment 8-2

Geocode immigrant-run businesses to Pittsburgh streets

As the 2010 Census shows, immigrants are becoming one of the most salient indicators of growth and wealth in a region. By looking at the immigrants who live in a city and analyzing where they decide to set up their businesses, city planners can investigate why certain neighborhoods are more immigrant-friendly than others, and turn their focus on the discovered qualities that make a neighborhood open and diverse. In this assignment you geocode and study immigrant-run high-tech firms, restaurants, and grocery stores in Pittsburgh.

Get set up

- Rename the folder \EsriPress\GIST1\MyAssignments\Chapter8\Assignment8-2YourName\ to your name or student ID. Store all files that you produce for this assignment in this folder.
- Create a new map document called **Assignment8-2YourName.mxd** with relative paths. Use a state plane projection.
- Create a new file geodatabase called **Assignment8-2YourName.gdb**. Save all new files that you create in this file geodatabase.

Build the map

Add the following spatial data to your map document:

- \EsriPress\GIST1\Data\Pittsburgh\City.gdb\ImmigrantBusinesses—sample of immigrant-run businesses in Pittsburgh. Attributes include: street address, city, state, ZIP Code, and type of business (Firm, Grocery, and Restaurant).
- \EsriPress\GIST1\Data\Pittsburgh\City.gdb\PghStreets—line features of Pittsburgh street centerlines.
- \EsriPress\GIST1\Data\Pittsburgh\City.gdb\Neighborhoods—polygon features of Pittsburgh neighborhoods.

Geocode immigrant businesses:

- Create an address locator, called **BusinessStreetsLocator**, to use when geocoding immigrant-run businesses to streets.
- Use US Address—Dual Ranges as the type of locator and PghStreets as the reference data. As address input fields, use both ADDRESS and ZIP.
- Save the geocoded businesses as **ImmigrantBusinesses**.

Rematch unmatched ZIP Codes:

You should get about 30 percent that do not match. Reasons include wrong ZIP Code data, place-names instead of street addresses, and incorrect or misspelled street address.

- Open the Interactive Rematch window and show results for Unmatched Addresses. The address, 700 Technology Dr, 15230, should be the first unmatched address listed. Start with it and work your way down the list until you get five more good matches.

- Use Internet sites such as https://www.usps.com/ or https://maps.google.com/ to rematch five unmatched addresses.

- When you find potential corrections, try making changes in the lower left panel of the Interactive Rematch window. If you get a good match, click the Match button. If you don't, skip that address and go to the next in the list.

- Keep a log of steps you took that found successful corrections and turn this in with your assignment as **Assignment8-2YourName.docx**. For each address investigated, give the original address, its problem, source for additional information, and correction. Provide the final match statistics in the log.

Requirements

- Create a map layout: Include a layout showing a point map of immigrant-run business locations in Pittsburgh. Symbolize the point map using type of business.

Assignment 8-3

Examine match option parameters for geocoding

There are no guidelines for selecting the match parameters that tune the ArcMap geocoding system. So here you explore the sensitivity of geocoding accuracy to variations in those parameters, and in the process make your own rough guidelines. You'll see that as match options become too lenient ArcMap starts making obvious errors in geocoding, making matches that are clearly wrong. The data is the same as for assignment 8-2—immigrant-run businesses for high-tech firms, restaurants, and grocery stores in Pittsburgh.

Get set up

- Rename the folder \EsriPress\GIST1\MyAssignments\Chapter8\Assignment8-3YourName\ to your name or student ID. Store all files that you produce for this assignment in this folder.
- Create a new map document called **Assignment8-3YourName.mxd** with relative paths. Use a state plane projection.
- Create a new file geodatabase called **Assignment8-3YourName.gdb**. Save all new files that you create in this file geodatabase.

Build the map

Add the following spatial data to your map document:

- \EsriPress\GIST1\Data\Pittsburgh\City.gdb\ImmigrantBusinesses—sample of immigrant-run businesses in Pittsburgh. Attributes include: street address, city, state, ZIP Code, and type of business (Firm, Grocery, and Restaurant).
- \EsriPress\GIST1\Data\Pittsburgh\City.gdb\PghStreets—line features of Pittsburgh street centerlines.
- \EsriPress\GIST1\Data\Pittsburgh\City.gdb\Neighborhoods—polygon features of Pittsburgh neighborhoods.

Geocode with alternative locators

- Create a locator called **PittsburghTIGER**. Use US Address–Dual Ranges as the type of locator and PghStreets as the reference data. As address input fields, use both ADDRESS and ZIP.
- Geocode the immigrant businesses data three times with different match parameter values. Click the Geocoding Options button in the Geocode Addresses window to change parameter values.
  - Strictest: Spelling sensitivity 100, Minimum candidate score 100, Minimum match score 100.

- Default: Spelling sensitivity 80, Minimum candidate score 10, Minimum match score 85. These are the locator's default matching parameters.
- Most Lenient: Spelling sensitivity 0, Minimum candidate score 0, Minimum match score 0.

Analyze results

Create a Word document, **Assignmemnt8-3YourName.docx**. Include a title and your name. Create and fill out the following table and include it in your document:

	Strictest		Default		Most Lenient	
	Number	**%**	**Number**	**%**	**Number**	**%**
Match						
Tied						
Unmatched						
Total	112		112		112	

Using your mapped points, first turn on the Strictest and Default point layers. Select Default points on the map that are not matched in the Strictest layer. Examine the selected Default records of each such point and type comparison address data for all of Pittsburgh in a table such as the one below. Include your table (and the next one) in Assignment8-3YourName.docx. The Good or Bad Match and Reason explains geocoding behavior in transitioning from the strictest to the default locator.

Strictest to Default

Raw Data Address	Raw Data Zip Code	Matched Address	Matched ZIP Code	Good or Bad Match and Reason

Do the same for the Default and Most Lenient layers. Zoom in to the Central Business District neighborhood and examine only points in that neighborhood (to save time) in a table such as the following:

Default to Most Lenient

Raw Data Address	Raw Data Zip Code	Matched Address	Matched ZIP Code	Good or Bad Match and Reason

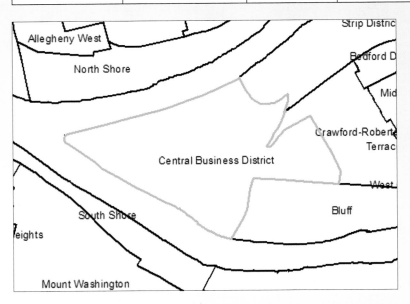

Make a recommendation on which locator to use and why for the three locators that you tried. If you were to try additional locators, which ones would you try next?

Part III
Analyzing spatial data

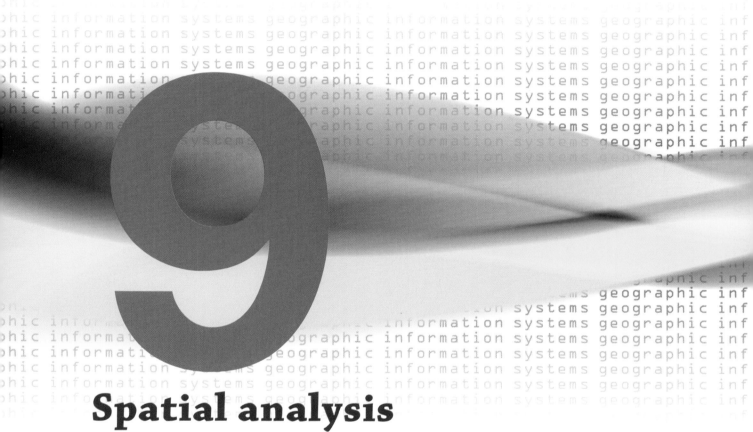

9
Spatial analysis

Analysis produces information from data that isn't always obvious, but is there. *Spatial analysis* deals with problems or behaviors that involve geographic locations. To produce sophisticated spatial information, you have to use methods or tools that process or transform data into information. This chapter has you apply several GIS tools for spatial analysis: buffers, spatial joins, intersection of map layers, and cluster analysis.

Several exercises in the chapter employ proximity analysis—determining what features are near other features. Of course, it is uniquely GIS that can perform such analysis with ease because map features have geographic coordinates. The tool for proximity analysis is a buffer, polygons that place an area around a given set of features. In another kind of problem, intersection of map layers allows you to find areas that meet criteria. In the facility location example of this chapter, you use buffers to represent criteria for locating new facilities. Areas that are the intersection of all criteria buffers are candidates for new facilities. One additional application of buffers is to estimate the falloff of facility use—public swimming pools in this case—the farther away the potential clients are from the nearest swimming pool. This is an example of the gravity model as used in GIS; the farther two objects are away from each other, the less attraction between them.

Spatial joins enable you to easily aggregate point data to polygon summaries. In the public pools case, the polygons for data summaries are buffers around pools. The results of this study can be used to identify gaps in the coverage of the target population by the set of public swimming pools in a city.

Finally, cluster analysis, while having a long history of development and use, is a major tool of the new field of data mining, searching for interesting and useful patterns in large, complicated datasets. You will search for patterns in tornadoes that touched down in the United States since 2000.

9-1
9-2
9-3
9-4
A9-1
A9-2
A9-3

Learning objectives

- *Buffer points for proximity analysis*
- *Conduct a site suitability analysis*
- *Use multiple ring buffers for calibrating a gravity model*
- *Use data mining with cluster analysis*

Tutorial 9-1

Buffering points for proximity analysis

Drug-free school zones are areas within 1,000 feet of a school that are designed to protect students from illicit drug dealers. Federal, state, and local laws increase penalties for illegal drug dealing within such zones. For example, Pennsylvania has a mandatory two-year minimum jail or prison sentence and a maximum four-year sentence for adult drug dealers in such cases. In this tutorial, you find all 911 drug calls for service to police for incidents in drug-free school zones. The output maps and table of incidents are materials that an analyst would study further—matching crime incident records and verifying distances from school properties to determine which cases are eligible for prosecution.

Open a map document

1 Start ArcMap and open Tutorial9-1.mxd from the Maps folder. This map is for the Middle Hill neighborhood of Pittsburgh, showing all schools in or near the neighborhood along with all drug calls for service to police during a certain period of time.

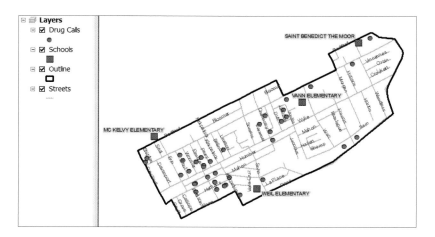

2 Save the map document to the Chapter9 folder of MyExercises.

Buffer schools

1 On the Menu bar, click Windows > Search.

2 In the Search window, type **buffer** in the search textbox and press ENTER.

3 Click Buffer (Analysis). This tool creates circular buffers (polygons) around points of a radius that you specify. It also creates buffer areas around and including the shapes of line or polygon features.

4 Type or make selections as shown in the image. The option of NONE for dissolve type has the buffer tool create a separate circular buffer for each school. Each finished buffer has the school's name, so you can intersect the buffers with drug calls and the calls within a buffer get the corresponding school name as an attribute—exactly what you need.

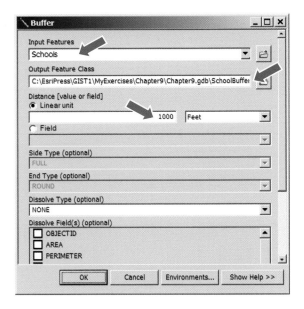

5 Click OK, wait for ArcMap to finish processing, and hide the Search window.

6 Move SchoolBuffers to the top of the table of contents and change its symbology to a hollow fill. See image on next page. You can see that a lot of the drug calls are in drug-free zones.

The next major step is to intersect the buffers and drug calls to identify the calls within buffers. First, however, you do some setup work to clean up attribute tables to display only needed attributes.

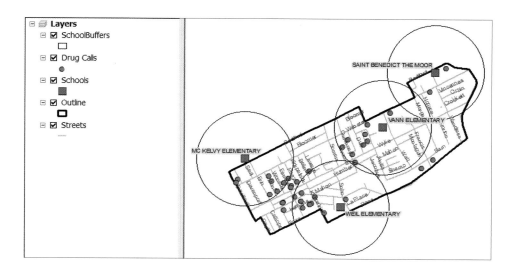

Turn off unneeded attributes

1 Open the property sheet's Fields tab for SchoolBuffers. There are 21 attributes, but you only need DESCR, which is actually the school name, and STUDENTS, which is enrollment level.

2 Click the Turn Off All Fields button at the top left of the window, click DESCR and STUDENTS to turn those fields back on. Click DESCR to select it and type **School** for its alias in the Appearance panel. Do the same for STUDENTS, giving it the alias **Enrollment**.

3 Click OK.

4 Similarly, for Drug Calls, display only NATURE_COD, ADDRESS, and CALLDATE with the aliases **Call Type**, **Address**, and **Call Date** respectively.

5 Save your map document.

Intersect school buffers and drug calls

1 Click Geoprocessing > ArcToolbox.

2 Expand Analysis Tools > Overlay, double-click the Intersect tool, and make selections or type as shown.

3 Click OK. Close ArcToolbox. The intersection point layer has just the Drug Call points inside school buffers.

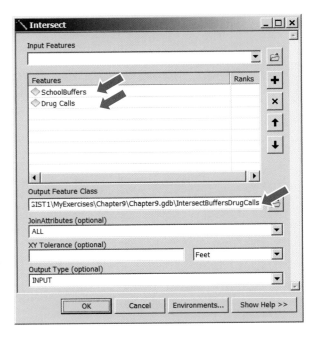

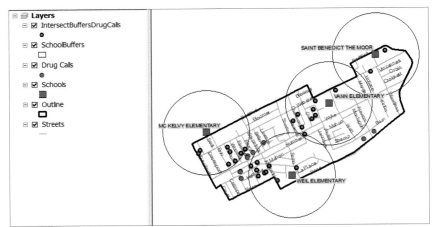

4 Open the attribute table for IntersectBuffersDrugCalls. You can see that each drug call has the name of the school corresponding to the buffer in which it lies. Also, the only attributes in the intersection table are the ones you left visible in the input map layers, and the intersection layer keeps the aliases you created for the input layers. In practice, you could export the table to Excel and give it to an analyst for further work.

IntersectBuffersDrugCalls					
School	Enrollment	FID_CADCalls	Call Type	Address	Call Date
WEIL ELEMENTARY	467	258	DRUGS	350 KIRKPAT	7/28/2012
WEIL ELEMENTARY	467	248	DRUGS	2140 HEMAN	7/24/2012
WEIL ELEMENTARY	467	246	DRUGS	2158 HEMAN	7/8/2012
WEIL ELEMENTARY	467	242	DRUGS	2162 HEMAN	8/2/2012
WEIL ELEMENTARY	467	223	DRUGS	2154 CENTRE	8/31/2012
WEIL ELEMENTARY	467	224	DRUGS	2154 CENTRE	8/20/2012

5 Close the table and save your map document.

> ### *YOUR TURN*
>
> You can buffer lines as well as polygons. Wylie Avenue is a major street with many commercial land uses that runs through the middle of the Middle Hill neighborhood. Commercial areas are ideal for illicit drug dealing, so an application of a buffer is to see how many drug calls are associated with Wylie Avenue. Use Select By Attributes to run the query "FNAME" = 'Wylie' to select all segments of that street. Then run the buffer tool with Streets as the input feature (ArcMap uses only the selected streets) with a 300-foot buffer and ALL Dissolve Type (the interior of individual line segment buffers are dissolved to make one polygon for this buffer). Save the buffer as MyExercises\Chapter9\Chapter9.gdb\WylieStreetBuffer. Turn off SchoolBuffers and IntersectBuffersDrugCalls. Symbolize the WylieStreetBuffer with a hollow fill. Save your map document.

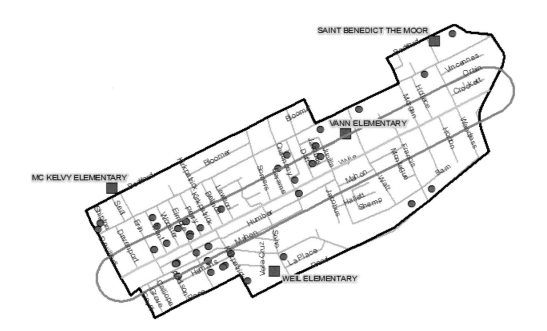

Tutorial 9-2

Conducting a site suitability analysis

Suitability analysis for facility location is a classic GIS application that depends on two GIS functions: buffers and intersection. In this tutorial, you find potential areas for new police satellite stations in each car beat of the Rochester, New York's Lake Precinct. A car beat is the territory of a patrol car. Criteria for locating these stations are that the site must be centrally located in each car beat (within a 0.33-mile radius buffer of car beat centroids), in retail/commercial areas (within 0.10 mile of a least one retail business), and within 0.05 mile of major streets. The suitable areas are those in the intersection of all three buffers.

Open a map document

1 Open Tutorial9-2.mxd from the Maps folder. This map document contains a map of the Lake Precinct of the Rochester Police Department. It includes police car beats, car beat centroids, retail business points, and street centerlines.

2 Save the map document to the Chapter9 folder of My Exercises.

You start out by making three sets of buffers for implementing siting criteria.

Buffer car beat centroids

1 Click Windows > Search, search for buffer, and open the buffer tool.

2 Type or make selections as shown in the image.

3 Click OK and hide the Search window.

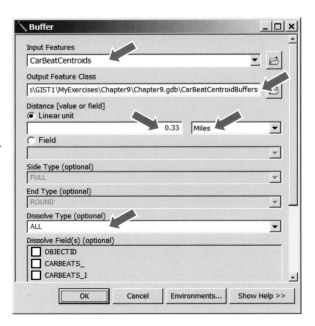

4 Symbolize the buffers polygons with a hollow fill, an outline color of Mars Red, and an outline width of 1.

Next, you need to find areas within the car beat buffers that meet the remaining criteria.

YOUR TURN

Create two more sets of buffers with the following characteristics:

Retail Businesses:

- Input Features = Retail Businesses
- Output Feature Class = RetailBusinessesBuffer (in MyExercises\Chapter9\)
- Linear unit = 0.10 miles
- Dissolve Type = ALL
- Symbolization is a hollow fill with a blue outline.

Streets:

- Use Select By Attributes to select Streets with the query condition, ""FCC" = 'A40' OR "FCC" = 'A41'". Use the Get Unique Values button for values of FCC.
- Input Features = Streets
- Output Feature Class = MajorStreetsBuffer
- Linear unit = 0.05 miles
- Dissolve Type = ALL
- Symbolization is a hollow fill with a green outline

Turn off the buffer inputs, leaving just the buffers on, to get a look at the results thus far.

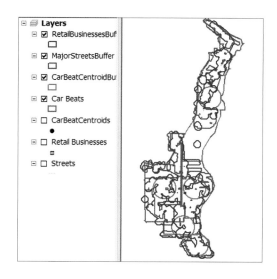

Intersect buffers

ArcMap lets you intersect only two map layers at a time. So the approach is to intersect two buffers and then to intersect the resulting intersected layer with the third buffer.

1 Search for **intersect** and open the Intersect (Analysis) tool

2 Type or make selections as shown.

3 Click OK.

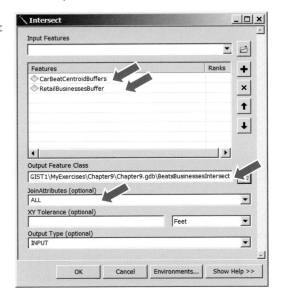

YOUR TURN

Intersect BeatsBusinessesIntersect with MajorStreetsBuffer to produce SuitableSites. Symbolize the end product as you wish and turn off the original buffers and Intermediate products, but turn Streets and Retail Businesses back on. The following map is zoomed in to one of the car beats to show some details. Save your map document.

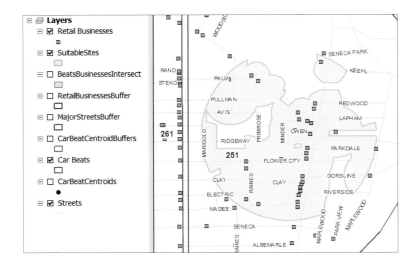

Tutorial 9-3

Using multiple ring buffers for calibrating a gravity model

The gravity model states that the farther two objects are from each other, the less the attraction between them. The falloff in attraction with distance is often non-linear and rapid, as in Newton's gravity model where the denominator of attraction is distance squared. Here you analyze the intention of youths, ages 5 to 17, to use public pools in Pittsburgh. During a budget crisis, Pittsburgh officials closed 16 of 32 public swimming pools. In the first year of the pools being closed, "pool tags," which have to be worn to gain admission to a pool, were free. So, if youths intended to use a pool, all they had to do was to get free pool tags. The question is whether distance from the nearest open pool had any impact on use.

A useful spinoff benefit of this case is that the available data is a random sample of pool tag holders so you get to deal with some issues associated with working with a random sample. Such a sample is representative of the total population, but of course is only a part of the population. For the pool case study, you need estimates for the total population of youths and therefore need to calculate a scale-up factor. There were 56,162 pool tag owners during the summer of the study year and the sample has 2,285 geocoded residences of pool tag owners. A separate sample estimates that 46.1 percent of pool tag holders are youths, ages 5 to 17. Thus, to scale the sample results of your analysis up to the population level, you need to multiply the number of sampled pool tag holders by a factor of 0.461 (56,162/2,285) = 11.3. The ratio in parentheses scales up to the total population of pool tag holders and the 0.461 adjusts that ratio to just youths 5 to 17. The 11.3 scale-up factor assumes that all parts of Pittsburgh have the same age distribution which, of course, leads to some errors.

Open a map document

1 Open Tutorial9-3.mxd from the Maps folder. The data for this tutorial includes the open and closed pools, the random sample of residence addresses of pool tag owners (youths in this case), and census block data on the population of youths. You use a multiple ring buffer to determine (1) the number of pool tag owners at a set of distances from the nearest open pool (scaled up from the sample by the 11.3 factor) and (2) the target youth population at the same set of distances. From this, you can calculate the percentage of youths who owned pool tags at a range of distances from the nearest open pool as well as the total youth population in each range.

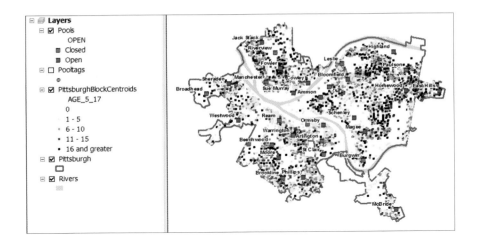

2 Save the map document to the Chapter9 folder of MyExercises.

3 Turn Pooltags on and then back off. Youths with pool tags live inside and outside of the city, as you can see.

You can't tell from the map whether youths closer to an open pool are more likely to have pool tags or not. You need to do some analytical work to answer that question.

Create multiple ring buffers for open pools

You are about to use the Multiple Ring Buffer tool in ArcToolbox. First, however, you have to select only open pools. Then the Multiple Ring Buffer tool creates buffers for only the open pools.

1 Click Selection > Select By Attributes, and type or make selections as shown.

2 Click OK.

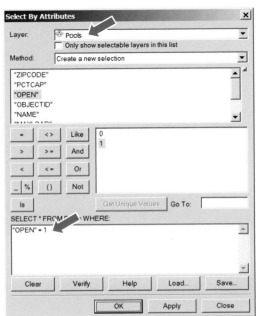

3 Search for and open the Multiple Ring Buffer (Analysis) tool. Type or make selections as shown. It doesn't matter which order you use to enter Distances. To match the figure, however, enter the "2" first and work backward to 0.25.

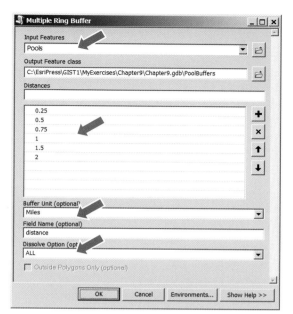

4 Click OK and close the Search window. Be patient. The Multiple Ring Buffer tool takes a minute or so to execute.

5 Click Selection > Clear Selected Features.

6 Open the Symbology tab of Layer Properties for PoolBuffers, click Categories, Unique Values, the Distance field, and Add All Values. Double-click the paint chip for the 0.25 row, select No Color for fill color, select 1.50 for outline width, Tourmaline Green (column 8, row 3) for outline color, and click OK.

7 Similarly change each successive paint chip with no color fill and outline colors working your way left from Tourmaline Green (Medium Apple, Peridot Green, Solar Yellow, Electron Gold, and Fire Red). The image shows the resulting map zoomed in to a part of Pittsburgh. Youths residing in areas

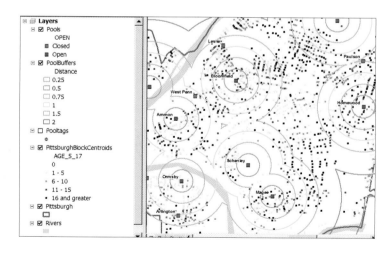

within green to yellow lines are fairly close to an open pool, within one mile. Those from yellow out to red are farther, from one to two miles. You can see there is a large area between orange and red lines (1.5 to 2.0 miles from the nearest pool) where there are no closed pools to reopen and fill the gap. Unless a new pool were built, that area looks to have poor access to public pools.

Spatially join buffers, pool tag owners, and youth population

You can easily use spatial joins to sum up the number of pool tag owners and youth population in each ring of the buffers and then calculate the corresponding use rate per buffer, the number of pool tag owners divided by youth population.

1 In the table of contents, right-click PoolBuffers, click Joins and Relates > Join and type or make selections as shown.

2 Click OK.

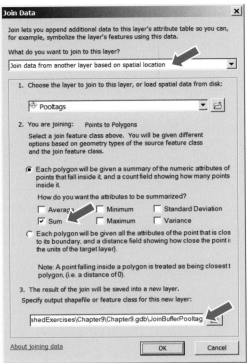

3 Open the attribute table of the new layer and see that it has the attribute Count_, which is the number of youths with pool tags in each ring of the buffer. Close the table.

4 Right-click JoinBufferPooltag, click Joins and Relates > Join, and type or make selections as shown.

5 Click OK.

6 Open the attribute table of the new layer and see that it has the attribute Sum_AGE_5_17, which is the number of youths in each ring of the buffer. Leave the table open.

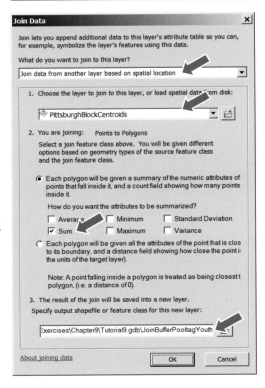

Compute use rate

1   Add a new field, UseRate, to the attribute table of JoinBufferPooltagYouth with the float data type.

2   Right-click UseRate, click Field Calculator.

3   Create the expression **UseRate = 100*11.3* [Count_] / [Sum_AGE_5_17]**. Click OK. Count_ is the number of records in each buffer ring, the same thing as the number of sampled pool tag owners. The 100 makes the computed fraction a percentage and the 11.3 is the scale-up factor from sample to population. Examine UseRate. It declines with distance as expected, from 87.4% in the 0- to 0.25-mile ring down to 41.0% in the 1.5- to 2.0-mile ring.

You will create a scatterplot of UseRate versus Distance, but first you need to compute a better distance for each buffer ring. The Distance attribute is the maximum radius of each buffer ring. Better representing each buffer is its average radius, which you compute next.

4   Add another new float data type field, DistanceAverage.

5   Click Customize > Toolbars > Editor.

6   Click the list arrow on the Editor toolbar > Start Editing, select JoinBufferPooltagYouth, and click OK> Continue.

7   Starting in the first row of the DistanceAverage column of the attribute table and working down the column, type the following values: **0.125, 0.375, 0.625, 0.875, 1.25, 1.75**. For this figure, columns were dragged left or right to group Distance, DistanceAverage, and UseRate so that you can see the relationship between these attributes.

Distance	DistanceAverage	UseRate
0.25	0.125	87.39027
0.5	0.375	65.25417
0.75	0.625	51.68109
1	0.875	44.07652
1.5	1.25	40.79899
2	1.75	41.04866

8   On the Editor toolbar, click the list arrow > Save Edits > Stop Editing, close the Editor toolbar, and close the attribute table.

Create a scatterplot of use rate versus average radius for each buffer ring

1   Click View > Graphs > Create Graph.

2   Type or make selections as follows:

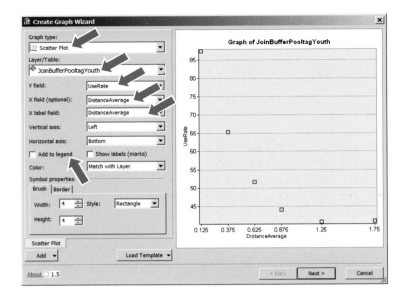

3 Click Next > Finish. You can see that the relationship between use rate and average buffer radius is smooth but with rapid falloff. Within 0.125 miles of an open pool, about 88 percent of youth have pool tags, but by 1.25 miles average distance, use rate is cut by more than half to 41 percent. The relationship includes some random error about a curve that you can imagine passing smoothly through the points due to having only a random sample plus assumptions made to allow use of the 11.3 scale-up factor.

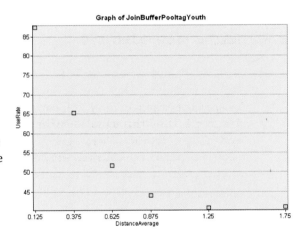

4 Save your map document and graph to the Chapter9 folder of MyExercises.

YOUR TURN

Turn off the two JoinBuffer map layers so that you can see the pools, your original buffers, PittsburghBlockCentroids (population by block of 5- to 17-year-old youths), and the Rivers. If you were director of parks in Pittsburgh and had the budget to open one more pool and also could close open pools and open closed pools on a one-for-one basis, what would you do? When finished, close your graph.

Tutorial 9-4

Using data mining with cluster analysis

Data mining uses exploratory methods to find patterns in large datasets—interesting or useful patterns that are hard to see by looking at data just in tables or even in graphs and maps. In this tutorial, you use a standard data mining method, k-means clustering, available in the Grouping Analysis tool of ArcToolbox. K-means partitions a set of n observations into k clusters. Each observation (or attribute table record) is assigned to the cluster with the closest mean for the attributes used for clustering. For example, in the following you cluster tornadoes into four groups in the United States using a scale for tornado severity (F-scale), number of injuries, and number of fatalities, and then look for geographic patterns on a map. K-means groups each tornado so that it is closest to its group mean for the attributes used compared to the other three groups.

Open a map document

1   Open Tutorial9-4.mxd from the Maps folder.

To keep the example manageable in terms of size, you will analyze tornadoes only from January 1, 2000 through December 31, 2008 (the latest date available) that have at least one injury or fatality. There were 11,645 tornadoes that touched down between 2000 and 2008, included in the Tornadoes2000 map layer, of which 912 had injuries or fatalities.

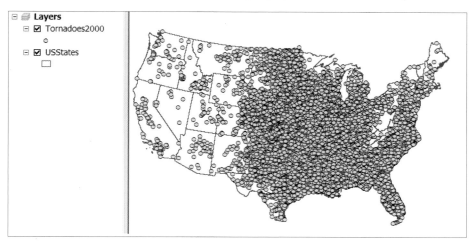

2   Open the Tornadoes2000 Properties and create the Definition Query, "INJ" > 0 OR "FATAL" > 0.

3 Open the attribute table of Tornadoes2000 and note the attribute TornadoID.

The tool that you use in the next exercise needs a unique ID for each record, but it cannot be a system-created ID such as FID (the first column of the table). So TornadoID was created as a new integer field and then given values using the Field Calculator by setting TornadoID = FID.

4 Close the attribute table and save the map document to the Chapter9 folder.

Run a cluster analysis

1 Search for and open the Grouping Analysis (Spatial Statistics) tool. You start out by clustering with only two of the three attributes of interest, one predictive of fatalities and injuries (F_SCALE) and the other a variable of interest (FATAL). The higher the F_SCALE value, the more severe the tornado. Then you cluster F_SCALE with INJ (number of injuries), and finally the tutorial has results for all three. It's often a good idea to take such an approach with clustering, looking at smaller numbers of attributes at first and then building up to more. You learn about behavior of phenomena as you go.

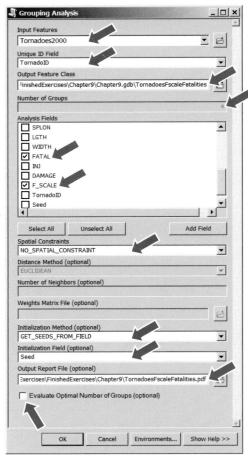

2 Type or make selections as shown. The k-means algorithm starts with an arbitrary four tornadoes to build four groups and then makes adjustments to the groups until good performance is achieved (meaning that variation within groups is minimized while variation across groups is maximized). You are running Grouping Analysis with the Initialization Method "GET_SEEDS_FROM_FIELD." The corresponding field in Tornadoes2000 is called Seed. It has the four tornadoes selected by the Grouping Analysis in a prior run with a different Initialization Method, "FIND_SEED_LOCATIONS," which you would normally use; however, every time the Grouping Analysis finds seeds, it finds different tornadoes and the resulting groups are different from run to run. So that your results match those in the figure at the bottom of the page, you will get the fixed four tornadoes from the Seed attribute.

3 **Click OK and close the Search window.** Be patient. It takes a minute or so for the Grouping Analysis to do its extensive calculations. When finished, the tool adds the clustered tornado points to the table of contents, symbolized with point markers of different colors.

Let's start analysis by examining the report that the tool produced.

4 **Use a Computer window to browse to the Chapter9 folder of MyExercises and open TornadoesFscaleFatalities.pdf.** Group numbers in your results may not match those in the following figure, but the values will. First, notice that the largest group has 695 tornadoes, the second largest has 175, the second smallest has 33, and the smallest has 9. So the largest group must be quite ordinary with so many tornadoes in it and the smallest group must be quite exceptional with so few. Next, examine the Variable-Wise Summary. The largest group has the lowest mean of fatalities (0.1885) and lowest F_SCALE (1.3640). The smallest group has the highest means for both fatalities (18.889) and F_SCALE (3.3333).

Next, you resymbolize TornadoesFscaleFatalities to show the order of severity more clearly and also to place the point markers for more severe cases on top of other points by changing the drawing order for the points.

Variable-Wise Summary

FATAL: R2 = 0.87

Group	Mean	Std. Dev.	Min	Max	Share	
1	0.1885	0.5122	0.0000	3.0000	0.1250	
2	18.8889	4.3319	13.0000	24.0000	0.4583	
3	6.4242	2.2834	4.0000	11.0000	0.2917	
4	0.5771	0.8774	0.0000	3.0000	0.1250	
Total	0.6732	2.3181	0.0000	24.0000	1.0000	

F_SCALE: R2 = 0.63

Group	Mean	Std. Dev.	Min	Max	Share	
1	1.3640	0.6470	0.0000	2.0000	0.4000	
2	3.3333	0.4714	3.0000	4.0000	0.2000	
3	3.2727	0.6639	2.0000	5.0000	0.6000	
4	3.1714	0.3769	3.0000	4.0000	0.2000	
Total	1.7993	0.9858	0.0000	5.0000	1.0000	

5 **Make a note of the group numbers for the most severe group (n=9), second most severe group (n=33), second least severe group (n=175), and least severe group (n=695).**

6 Leave your PDF report open for reference in the next exercise.

Resymbolize clustered points

1   Open the Symbology tab of Properties for TornadoesFscaleFatalities.

2   Click in the row for the most severe group (which has 9 members) and click the up arrow button until that group is the first. Continue moving rows until you have the order from the top running from most severe to least severe.

After step 6 that follows, this will be the new drawing order for tornados. The last to be drawn is the one with the fewest but most important points, so that it is on top and therefore visible. Next, you change color fill so that the colors range top down from red, to yellow, to green, and to blue to show order using the color spectrum.

3   Double-click the point marker symbol in the first row, change its color to Mars Red (column 2, row 3 of the color pallet).

4   Likewise, make the second-row point marker Solar Yellow (column 5, row 3), the third-row point marker Medium Apple (column 7, row 3), and the last point marker Big Sky Blue (column 9, row 3).

5   Type new values for labels as follows: red point marker becomes **Most Severe**, yellow point marker becomes **Second Most Severe**, green point marker becomes **Second Least Severe**, and blue point marker becomes **Least Severe**. Click OK.

6   In the Symbology tab, click the Advanced button > Symbol Levels > the check box for "Draw this layer using the symbol levels specified below" > OK > OK. Now everything of importance is visible and the map is much easier to read. There appears to be a pattern of more severe tornadoes in a swath as indicated on the map. Most affected are the lower right corner of Kansas and the areas bordering Missouri and Arkansas, and Kentucky and Tennessee.

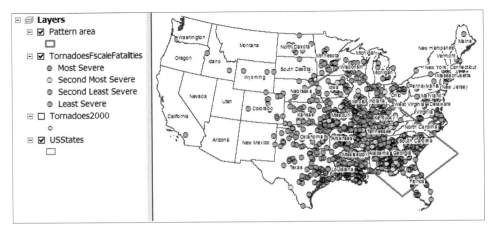

7   Save your map document.

YOUR TURN

Run a second cluster analysis with the Group Analysis tool with the same input and settings except use F_SCALE and INJ as the two attributes to cluster and name the new layer TornadoesFscaleInjuries and the report TornadoesFscaleInjuries. Study the report and resymbolize the group point markers as you did above. The following are the finished map, and then the map for running all three attributes in the cluster analysis. As you might expect, injuries follow a similar geographic pattern, as do fatalities. The picture of the most severely impacted states changes a bit when considering both fatalities and injuries, but clearly, it's the lower right corner of Kansas, the areas bordering Missouri and Arkansas, and Kentucky and Tennessee that are most severe. When finished, save your map document and close ArcMap.

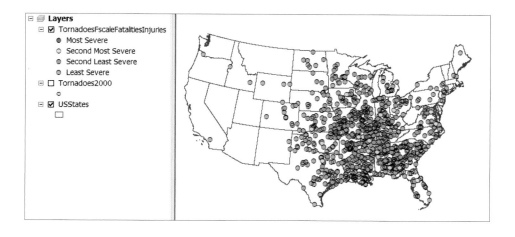

Assignment 9-1

Analyze population in California cities at risk for earthquakes

In this assignment, you use GIS to create buffers around major earthquakes that have occurred in California and determine how many people live in urban areas near these events. In the process, you learn a few new ways to work with the Buffer tool.

Get started

First, rename your assignment folder, create a map document, and create a file geodatabase.

- Rename the folder \EsriPress\GIST1\MyAssignments\Chapter9\Assignment9-1YourName\ to your name or student ID. Store all files that you produce for this assignment in this folder.
- Create a new map document, **Assignment9-1YourName.mxd** with relative paths.
- Create a new file geodatabase, **Assignment9-1YourName.gdb**. Save all of the map layers and tables you create in this file geodatabase.

Requirements

Use a UTM projection appropriate for California in your data frame. Use this projection for all map layers that you create. Include the following map layers and data in your map document:

- \EsriPress\GIST1\Data\UnitedStates.gdb\CACounties—polygon features of California counties
- \EsriPress\GIST1\Data\UnitedStates.gdb\USCities_dtl—point features for cities in the United States. Attribute POP2007 has city populations. Records without data for this attribute have the value -99999, which stands for missing value. Use a query definition to eliminate those cities with a value of -99999. Create a new feature class, CACities, with only California cities and remove the original USCities_dtl layer.
- \EsriPress\GIST1\Data\DataFiles\Earthquakes.dbf—table of earthquakes in California, off the coast of California, and in southern Oregon with latitude and longitude attributes in North American Datum 1983 coordinates. Be sure to specify these coordinates, and not UTM, when using Create Feature Class From XY Table ArcCatalog. The attribute MMI—the Modified Mercalli Intensity scale, currently used in the US—has values reflecting expected impacts. Values 7 to 12 correspond to increasing damage, from considerable damage of poorly constructed buildings through total destruction.

Create a 20-mile buffer around earthquakes whose MMI is 7 or greater for 1985 through to the end of the data. Include a label with the total population within buffers. Export a layout with your map, **Assignment9-1YourName.jpg**.

It takes a couple of steps to get the desired buffers for this assignment. The map needs each separate buffer area to have a label displaying the total urban population in that area. If you were to

9-1
9-2
9-3
9-4
A9-1
A9-2
A9-3

use the ALL dissolve type when buffering earthquakes, a single polygon would result for all buffers, even though they are separate areas. Some 20-mile buffers overlap. Each set of overlapping buffers needs to be dissolved to form a single buffer polygon. Many 20-mile buffers do not overlap. These need to be separate polygons. The approach to building the needed buffers uses a dissolve field that you can create in the following steps:

- Buffer earthquakes with an MMI greater than or equal to 7 from 1985 and later using a 20-mile radius, the NONE dissolve type, and saved in the feature class Earthquakes_Buffer. Do not use the dissolve fields in the buffer tool, although you have to do some dissolving in this assignment later. Several circular buffers overlap.

- Open the attribute table of Earthquakes_Buffer and add a new field called **BufferGroup** with the Short Integer data type.

- Work from north to south. Use the Select Features tool to select the overlapping buffers along the north coast of California. Then, using the Field Calculator or Editor toolbar, set the members of the selected buffers to the value 1 in the BufferGroup field. Repeat this step for the next set of overlapping buffers but use the value 2 for BufferGroup. Finally, give the remaining two groups of overlapping buffers unique values for BufferGroup.

- Use the Dissolve tool to dissolve Earthquakes_Buffer using BufferGroup as the dissolve field. Don't use the statistics fields in the dissolve tool. Call the new feature class Buffer _Earthquakes2, and save it to your file geodatabase. The end result is separate polygon buffers for each non-overlapping buffer plus polygons for each set of overlapping buffers.

- Spatially join CACities to Buffer_Earthquakes2 to create Join_CitiesBuffer. Start the join by right-clicking Buffer_Earthquakes2, and be sure to use SUM so that city attributes are summed by buffer polygon for labeling your map. Use SUM_POP_2007 to label the buffers.

- Rename and symbolize the layers in the table of contents.

Assignment 9-2

Analyze visits to the Jack Stack public pool in Pittsburgh

Tutorial 9-3 had you analyze intentions of youths (or their parents/guardians) to use Pittsburgh public pools during a specific summer. Intention was represented by owning a pool tag needed for admission to a pool. You found that the intended use rate, calculated as the percentage of youths who owned pool tags, fell off quickly with distance from the nearest open pool. In this assignment you study actual visits of youths to one particular public pool in Pittsburgh, called Jack Stack. A limitation of your study is that you do not determine how many pool tag owners near enough to Jack Stack pool never visited it. You only analyze pool tag owners who visited the pool at least once. (While not a requirement of this assignment, you could analyze the pool tag data and see if you can determine how many pool tag owners near Jack Stack never visited the pool. You'd find that most visited at least once.)

Get started

First, rename your assignment folder and create a map document.

- Rename the folder \EsriPress\GIST1\MyAssignments\Chapter9\Assignment9-2YourName\ to your name or student ID. Store all files that you produce for this assignment in this folder.

- As a shortcut to building your map document, open Tutorial9-3 from the Maps folder and save it as \EsriPress\GIST1\MyAssignments\Chapter9\Assignment9-2YourName\ **Assignment9-2YourName.mxd** where you use your actual name or student ID. Remove the Pooltags layer from the map document.

Process the data

Add \EsriPress\GIST1\Data\Pittsburgh\PublicPools.gdb\JackStackVisits to your map document. This is data from a random sample of visitors to Jack Stack pool. Attributes include:

- POOL = name of the pool that the pool tag owner stated that he/she intended to use
- NOPERSONS = number of persons in the party or family for a visit
- AGE = age of the person in the party included in the random sample
- VISITS = number of visits that the person included in the random sample actually made to the pool during the summer studied

Requirements

- Add the following definition query to JackStackVisits so that you analyze only youths:
  "AGE" <=17 AND "AGE" >= 5

- Create a multiple ring buffer layer for the JackStack pool called **JackStackBuffer** using the same buffer distances as in Tutorial9-3: 0.25, 0.50, 0.75, 1.00, 1.50, and 2.00 miles. **Hint:** Select Jack Stack pool in the Pools layer to restrict the buffer to that pool. Intersect the buffer with Pittsburgh in a layer called **IntersectPittsburghBuffer** and use this intersection layer or further joins to it for display and spatial joins. You need to exclude pool users residing outside of Pittsburgh, and the intersection accomplishes that. While non-Pittsburgh youths are welcome to use Pittsburgh public pools, as a matter of policy the pool system evaluation is limited to Pittsburgh residents.

• Compute the following average for VISITS for each buffer ring. ***Hint:*** Right-click IntersectPittsburghBuffer, click Joins and Relates > Join, select Join data from another layer based on spatial location, join JackStackVisits, and select Average. Call this feature class **JoinBufferAverageVisits**.

• Add an attribute, called DistanceAverage, to JoinBufferAverageVisits for average distance of buffers as done in Tutorial 9-3. Use the Edit toolbar, edit the attribute table, and type the distances **0.125**, **0.375**, **0.625**, **0.875**, **1.25**, and **1.75**, starting in the top row. There is no need to use the 11.3 scale-up factor from Tutorial 9-3. The average that you calculate is representative of the population.

• Create a scatterplot of average number of visits by youth versus average distance to buffers and include it in a layout with your map. Add the graph to your layout by right-clicking it. While the relationship has more scatter than the one you found for intentions, evident is a rapid decline in average visits with increasing distance from the pool.

Assignment 9-3

Use data mining with crime data

Larceny is the crime committed when one person takes the personal property of an owner with the intent of depriving the owner of it. Let's see what you can learn about larceny crimes in Pittsburgh by analyzing them with cluster analysis.

Get started

First, rename your assignment folder and create a map document.

- Rename the folder \EsriPress\GIST1\MyAssignments\Chapter9\Assignment9-3YourName\ to your name or student ID. Store all files that you produce for this assignment in this folder.
- Create a new map document called **Assignment9-3YourName.mxd** with relative paths.

Process the data

Add the following to your map document:

- \EsriPress\GIST1\Data\Pittsburgh\City.gdb\Neighborhoods—polygon features of Pittsburgh neighborhoods. Symbolize it with a hollow fill and label it with the neighborhood name, the attribute HOOD.
- \EsriPress\GIST1\Data\Pittsburgh\Shapefiles\Larcenies.shp—point features with a year's worth of larcenies in Pittsburgh. Records are included, however, only where there was an arrest, so that data is included on the perpetrator. Attributes of interest are:

ARRAge = age of the arrested person at the time of arrest. Included are juveniles and adults.

Gender = 0 for females and 1 for males

Race = 0 for non-whites and 1 for whites

Seed is the column for initiating a cluster analysis with six clusters.

Requirements

- Run a cluster analysis (Group Analysis tool) using Larcenies as the input, OBJECTID_1 as the unique ID field, 6 groups, No_SPATIAL_CONSTRAINT, and GET_SEEDS_FROM_FIELDS (Seed). Include three attributes: ARRAge, Gender, and Race. Output a report.
- Give your groups descriptive names in the groups legend, for example, "White Males 40s."
- The key to making sense of cluster groups on a map is good symbolization. Study the groups that result from the cluster analysis and use similar colors for related groups. For example, use bright red for white males in their 40s and a light red for adult white males in their 20s.

Study question

What patterns can you make out for four parts of Pittsburgh, the three parts separated by rivers (call them North, South, and Central) as well as the Central Business District (neighborhood to the right of where the three rivers join)? State what you think is the strongest or most interesting spatial pattern, in terms of the six clustered groups for each of the four geographical areas, using visual observation. Include a zoomed in map for each area from a layout. Include the maps and observations in your document **Assignment9-3YourName.docx**.

Part III
Analyzing spatial data

ArcGIS 3D Analyst for Desktop

This chapter is an introduction to the ArcGIS 3D Analyst extension to ArcGIS for Desktop that enables 3D display and processing of maps. 3D viewing can provide insights that would not be readily apparent from a 2D map of the same data. For example, instead of inferring the presence of a valley from 2D contours, in 3D you actually see the valley and the difference in height between the valley floor and a ridge. This chapter uses topography, curb, and building data from the city of Pittsburgh's Mount Washington and Central Business District neighborhoods to show you how to display and analyze data in 3D. It also introduces ArcGlobe, a web service from Esri based on 3D that includes rich basemaps.

Learning objectives

- *Create 3D scenes*
- *Create a triangulated irregular net-work (TIN)*
- *Drape features over a TIN*
- *Navigate through 3D scenes*
- *Create a 3D animation*
- *Use 3D effects*
- *Edit 3D objects*
- *Conduct landform analysis*
- *Explore ArcGlobe*

Tutorial 10-1

Creating a 3D scene

ArcGIS 3D Analyst is one of a collection of extensions to the basic ArcGIS for Desktop software package. You must have the extension installed on your computer and then enable it in ArcMap, which you will do next.

Add the ArcGIS 3D Analyst extension and toolbar

1 Start ArcMap with a new blank map.

2 On the Menu Bar, click Customize > Extensions.

3 Select the check box beside the 3D Analyst extension > Close.

4 Click Customize > Toolbars > 3D Analyst. The toolbar for 3D Analyst appears. You can dock it with other toolbars below the main menu.

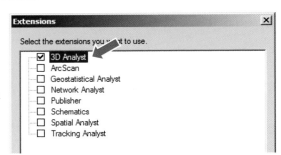

Launch ArcScene and add a topography layer

ArcMap is for 2D maps, so to work with 3D maps you need a new package. It's called ArcScene.

1 On the 3D Analyst toolbar, click the ArcScene button.

10-1
10-2
10-3
10-4
10-5
10-6
10-7
10-8
10-9
10-10
A10-1
A10-2
A10-3

2 Click New Scenes > Blank Scene > OK. A new, untitled scene window opens.

3 In ArcScene, click the Add Data button, browse through the Data folder to 3DAnalyst.gdb, and click Topo > Add. A topography layer of contours near downtown Pittsburgh appears as a 3D view, although the display is 2D at this time.

You make it 3D next.

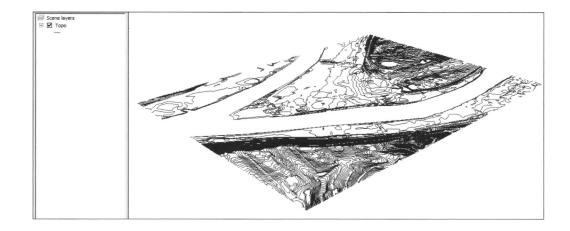

Set document properties and save the 3D scene

1 Click File > Scene Document Properties.

2 Click the Pathnames Option button to store relative paths > OK.

3 Click File > Save As, navigate to the Chapter10 folder of MyExercises, and save the 3D scene as **Tutorial10-1.sxd**. ArcScenes have a different file type—SXD—from map documents.

Tutorial 10-2

Creating a TIN (triangulated irregular network) from contours

3D uses the TIN representation for modeling surfaces. TIN is a vector data model of contiguous, non-overlapping triangles with vertices created from adjacent sample points of x, y, and z values from 3D space. For example, next you create a TIN from the topography contour map.

Create a TIN

1 Save the scene as **Tutorial10-2.sxd** to the Chapter10 folder.

2 Click the ArcToolbox button, click the plus (+) sign beside 3D Analyst Tools, click the plus sign beside Data Management, and click the plus sign beside TIN.

3 Double-click Create TIN.

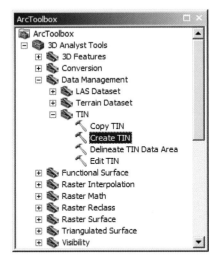

4 In the Create TIN window, save the output TIN as **chapter10_tin** to the Chapter10 folder of MyExercises.

5 Set Coordinate System to NAD_1983_StatePlane_Pennsylvania_South_FIPS_3702_Feet.

6 Select Topo as the input feature class and Hard_Line as the SF Type.

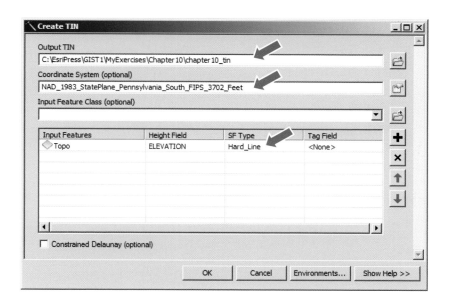

7 Click OK. Wait while ArcGIS 3D Analyst creates a TIN from the topography contour lines and adds it as a new layer.

8 Close ArcToolbox. The resulting map shows both the original topography (2D) and the TIN (3D).

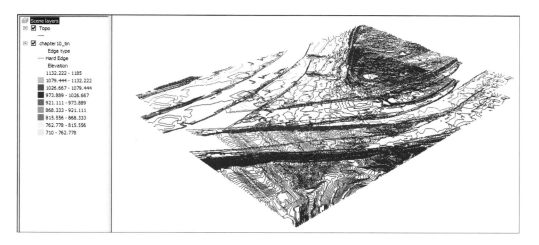

Change TIN appearance

1 In the table of contents, right-click chapter10_tin, and click Properties.

2 Click the Symbology tab and clear the check boxes beside Edge Types and Elevation to turn these off.

10-1
10-2
10-3
10-4
10-5
10-6
10-7
10-8
10-9
10-10
A10-1
A10-2
A10-3

3 Click the Add button, click "Faces with the same symbol", click Add, and close the Add Renderer window.

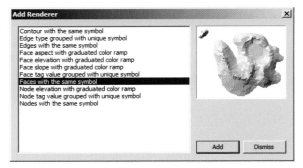

4 Click the Symbol color for Faces, click Early Desert Tan, and OK. This creates a more realistic looking surface map.

5 Click Apply and OK. The topography layer is now one color.

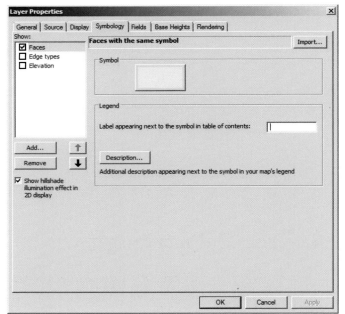

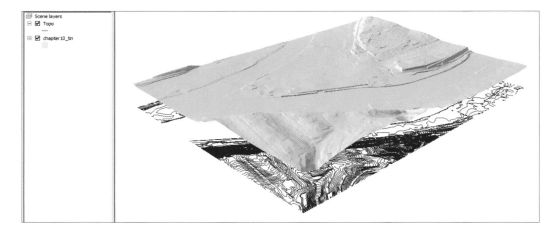

10-1
10-2
10-3
10-4
10-5
10-6
10-7
10-8
10-9
10-10
A10-1
A10-2
A10-3

Navigate a 3D view

You do not have a good view of the TIN yet, but you can get one with some navigation.

1 On the Tools toolbar, click the Navigate button.

2 Click and drag the map to view the scene from different angles. You see that Pittsburgh is fairly flat where the three rivers converge (this area is known as "the Point") and hilly in the Mount Washington neighborhood to the right of the Point in the image.

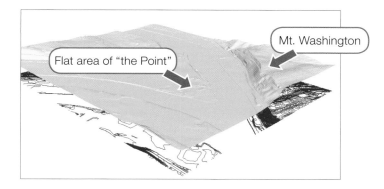

Zoom to see TIN's triangles

Zooming to a small area allows you to see the triangulated irregular network.

1 Click the Zoom In button.

2 Zoom to a small area on Mount Washington. There are many small triangular facets making up the surface of the chapter10_tin.

3 Click the Full Extent button.

4 Save the scene.

Tutorial 10-3

Draping features onto a TIN

Now that you have a 3D TIN, you can drape 2D map layers on it to show them in 3D.

Drape curbs

1 Save the scene as **Tutorial10-3.sxd** to the Chapter10 folder.

2 In the table of contents, turn the original Topo layer off.

3 Click the Add Data button, browse to 3DAnalyst.gdb, click Curbs > Add. Curbs appear below the TIN contours.

4 In the table of contents, right-click the Curbs layer and click Properties.

5 Click the Base Heights tab and select the radio button beside Floating on a custom surface. The base height is the elevation at which this flat map layer appears.

6 Type **10** as the Layer offset number. This elevates the sidewalks slightly so they do not "bleed" into the contours.

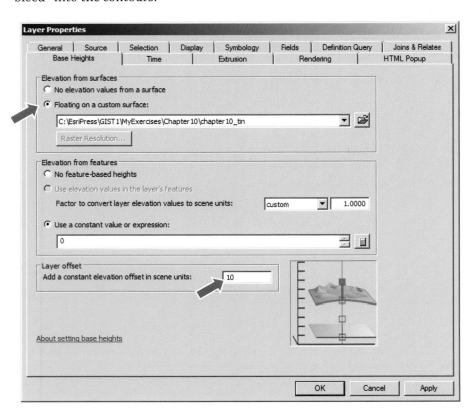

10-1
10-2
10-3
10-4
10-5
10-6
10-7
10-8
10-9
10-10
A10-1
A10-2
A10-3

7 Click Apply > OK.

8 Change the color of the Curbs layer to medium/dark gray. The resulting map displays the curbs draped over contours.

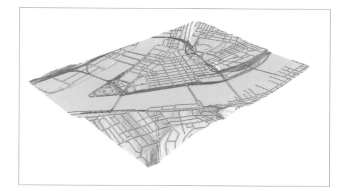

YOUR TURN

Add rivers polygon features from 3DAnalyst.gdb, change the color to Water Body blue, and drape it to the TIN. Set the Layer offset to 3. Use the Navigate tool to view the scene from different angles. When finished, zoom to full extent.

Drape and extrude buildings

1 Click the Add Data button, browse to 3DAnalyst.gdb, click Bldgs > Add. This action adds the buildings layer with number of building stories in the attribute table. The number of stories for Pittsburgh buildings is fictitious but a good estimate.

2 Right-click the Bldgs layer and click Properties.

3 Click the Base Heights tab, select the radio button beside Floating on a custom surface, and choose chapter10_tin.

4 Type **5** for the Layer offset.

5 Click the Extrusion tab and select the check box beside Extrude features in layer.

6 Enter the following expression for the Extrusion value. The building height is approximated assuming that each story is 10 feet high. You can use the calculator expression button to choose the field.

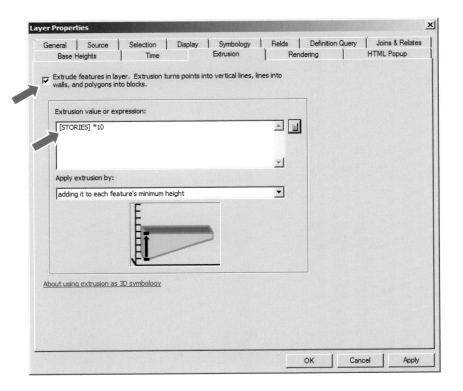

7 Click OK.

8 Change the color of the buildings to Sunset Orange. The resulting view is buildings with various heights all positioned just above the TIN. Most of the buildings in the Mount Washington neighborhood are residential houses, while the downtown area, of course, contains high-rises.

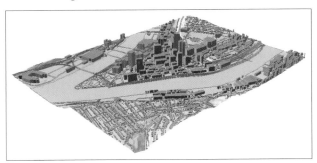

Drape an aerial image

In addition to vector features, you can drape images onto a TIN. In this exercise you will drape a USGS aerial image downloaded as a National Agriculture Imagery Program (four band) file. A better quality image could also be downloaded but would be a much larger file size.

1   Turn the Bldgs layer off and navigate to a view similar to the following:

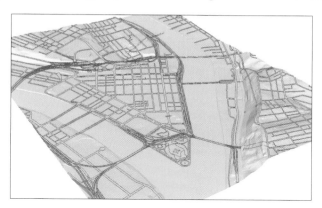

2   Click the Add Data button, browse to 3DAnalyst.gdb, click Image > Add. This action adds the USGS aerial image.

3   Right-click the Image layer, click Properties > Base Heights tab, and make the selections as follows and click OK. The USGS image uses UTM coordinates whose base unit is a meter, so some conversions are necessary. The USGS image will be draped to the TIN.

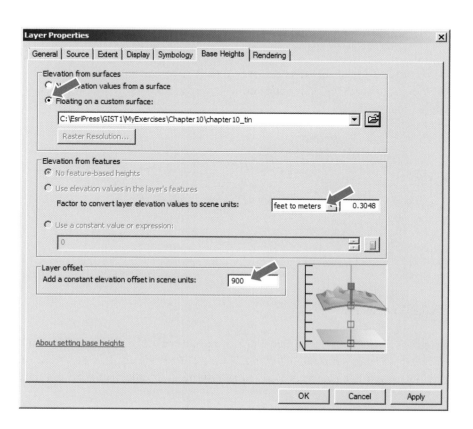

10-1
10-2
10-3
10-4
10-5
10-6
10-7
10-8
10-9
10-10
A10-1
A10-2
A10-3

YOUR TURN

Practice adjusting the base height properties of the USGS image. Save your scene.

10-1
10-2
10-3
10-4
10-5
10-6
10-7
10-8
10-9
10-10
A10-1
A10-2
A10-3

Tutorial 10-4

Navigating scenes

Navigation in 3D is more complicated than in 2D, so ArcScene adds additional navigation tools.

Center view on target location and set observer

1 Save the scene as **Tutorial10-4.sxd** to the Chapter10 folder.

2 Remove the Image layer, turn Bldgs on, and zoom to the map extents.

3 On the Tools toolbar, click the Center on Target button.

4 Click a location on the Mount Washington neighborhood that overlooks the city of Pittsburgh.

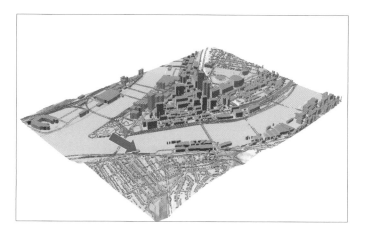

5 On the Tools toolbar, click the Set Observer button.

6 Click a location where the three rivers meet, called the Point, to set the observer location. The resulting view is from the perspective of an observer at the Point, looking toward the Mount Washington neighborhood.

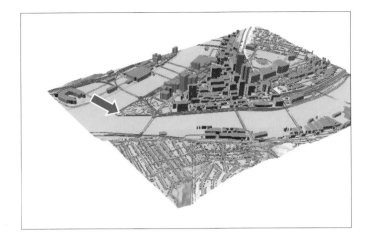

Fly through a scene

1 Zoom to the full extent.

2 Set the observer location to the confluence of the three rivers at the Point. The scene smoothly shifts to that location.

10-1
10-2
10-3
10-4
10-5
10-6
10-7
10-8
10-9
10-10
A10-1
A10-2
A10-3

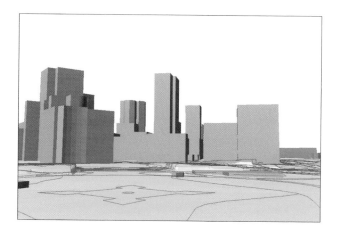

3 From the Tools toolbar, click the Fly button.

Get ready for a wild flight!

4 Click anywhere in the scene with the bird cursor, and click again to start your flight.

5 Slowly move the mouse to the left, right, up, or down. Click the left mouse button to increase your speed, and click the right mouse button to decrease your speed.

6 Press the Escape (ESC) key on the keyboard to stop the flight. A new view appears where you stopped the fly-through.

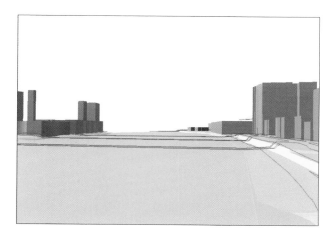

7 Zoom to the full extent.

Create multiple views

3D views can be complicated. Using multiple views allows you to zoom to small areas when still zooming to the overall map.

1 From the Standard toolbar, click the Add Viewer button 🖼 .

2 With the Viewer1 window selected, click the Target, Observer, Navigate, and Zoom buttons to change the view. You can also use the scroll wheel on your mouse to zoom in and out.

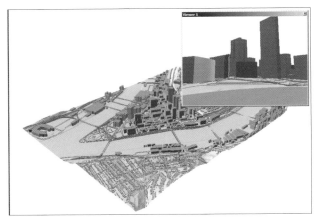

3 Save the scene.

Tutorial 10-5

Creating an animation

3D animations allow you to record movements within your views so that you can save and play them back at a later time.

10-1
10-2
10-3
10-4
10-5
10-6
10-7
10-8
10-9
10-10
A10-1
A10-2
A10-3

Add the animation toolbar

1   Save the scene as **Tutorial10-5.sxd** to the Chapter10 folder.

2   Close the Viewer1 window.

3   Right-click anywhere in the blank area of a toolbar.

4   Click Animation to display this toolbar. The Animation toolbar appears.

5   Zoom to the full extent.

Record and play an animation

1   Click the Open Animation Controls button on the Animation toolbar.

2   Click the Record button ⬤ .

3   Click the Fly button and create a fly-through anywhere in your scene, and then click the Esc key to end your flight.

4   Click the Stop button ■ .

5   Click the Play button ▶ .

Save an animation

1   On the Animation toolbar, click Animation > Save Animation File.

2   Save the animation as **Tutorial10-5** to the Chapter10 folder. This saves the animation as an ArcScene (.asa) file.

Load an animation

1 On the Animation toolbar, click Animation > Load Animation File.

2 In the Open Animation dialog box, navigate to your Chapter10 folder, click Tutorial10-5.asa > Open.

3 Click the Play button from the Animation Controls toolbar.

4 Zoom to the full extent.

> **YOUR TURN**
>
> Practice creating animations by zooming in to a small area first. Explore the Options menu in the Animation Controls toolbar to see animation play and restore options.

Export an animation to video

If you wish to share your animation with someone who does not have ArcScene, you can export it in a common format such as an .avi file.

1 On the Animation toolbar, click Animation > Export Animation.

2 Navigate to the Chapter10 folder, save your animation as **Tutorial10-5.avi**, and then click Export > OK. Wait until the animation is fully exported.

3 Close the Animation Controls and Animation toolbars and save your scene.

> **YOUR TURN**
>
> Launch a video player such as Windows Media Player and play the AVI video that you created. If you have trouble loading and playing your video, choose the animation file in the Chapter10 folder.

GIS TUTORIAL 1   **ArcGIS 3D Analyst for Desktop** **CHAPTER 10** 359

10-1
10-2
10-3
10-4
10-5
10-6
10-7
10-8
10-9
10-10
A10-1
A10-2
A10-3

Tutorial 10-6

Using 3D effects

Special effects such as transparencies, lighting, and shading modes can enhance the 3D experience for the viewer. For example, a city planner might want to differentiate between existing and proposed buildings by making one or the other transparent. In this tutorial you select buildings in one neighborhood to make them transparent and add proposed buildings as solid polygons.

The Layer Face Culling command turns off the display of front or back faces of an aerial feature or graphic. Layer Lighting turns lighting on or off for the selected layer. Shading Mode allows you to define the type of shading (smooth or flat) to use for the layer selected. Depth Priority allows you to define which 3D layer should be given higher priority. This is useful when you have two 3D polygon layers that share the same location and might obstruct each other, such as land parcels and buildings.

Prepare a study area

1 Save your scene as **Tutorial10-6.sxd** to the Chapter10 folder and zoom to full extent.

2 Choose Selection > Select By Attributes and build the query as shown. This selects buildings on the North Shore neighborhood.

3 In the table of contents, right-click Bldgs, click Selection > Create Layer from Selected Features.

4 Rename the layer **North Shore Bldgs** and clear the selected features.

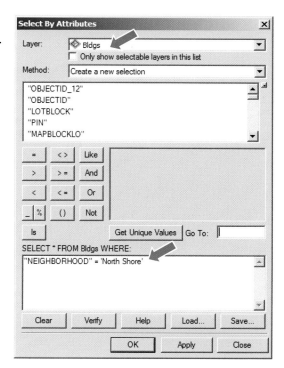

5 In the table of contents, right-click North Shore Bldgs, click Properties > the Base Heights tab.

6 Click "Floating on a custom surface" and type **10** for the Layer offset.

7 Click the Extrusion tab, select the check box beside Extrude features in a layer, enter the expression **[STORIES]** *****10**, and click OK.

8 Change the North Shore Bldgs color to Ultra Blue.

9 Turn off the original Bldgs layer and zoom to the North Shore Bldgs layer. The 3D scene shows the existing buildings on the North Shore.

> ### *YOUR TURN*
>
> Add the layer from 3DAnalyst.gdb called ProposedBldgs, drape them to the TIN, extrude using [STORIES] * 10, an offset from surface of 10, and a Mars Red color.

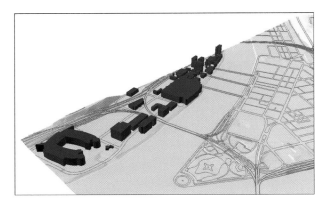

Create a transparency effect

1 Right-click anywhere in the blank area of a toolbar and click 3D Effects to display the 3D Effects toolbar.

2 On the 3D Effects toolbar, click the Layer drop-down, click North Shore Bldgs > the Layer Transparency button.

3 Drag the slider to change the layer's transparency to 75%.

4 On the 3D Effects toolbar, click the Layer Face Culling button 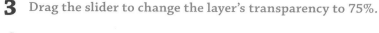, and select the radio button option beside Honor features built-in face culling. The resulting effect is a nice transparency for existing buildings on the North Shore to better focus on the proposed buildings.

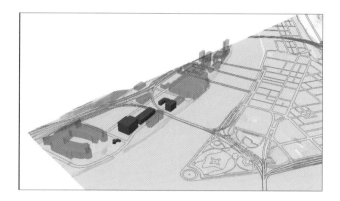

YOUR TURN

Experiment with changing other effects for your 3D buildings, including Layer Lighting, Shading Mode, and Depth Priority. Change the effects for the Rivers and Curbs layers in your 3D scene.

Tutorial 10-7

Using 3D symbols

*The 3D Analyst extension comes with many 3D symbols for objects such as trees,
which you will use next.*

Add trees layer

1 Save your scene as **Tutorial10-7.sxd** to the Chapter10 folder.

2 Click the Add Data button.

3 In the Add Data browser, browse to 3DAnalyst.gdb, click NorthShoreTrees > Add.

Display points as 3D trees

1 In the table of contents, right-click the NorthShoreTrees layer and click Properties.

2 Click the Base Heights tab and select the radio button beside Floating on a custom surface.

3 Click the Symbology tab and click the Symbol button.

4 In the Symbol Selector window, click the Style References button (in the lower right of the window), select 3D Trees, and click OK twice.

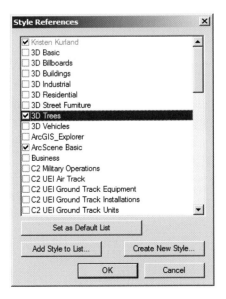

5 Click Unique Values under Categories, click TYPE for the value field, click Add All Values, and double-click the point symbol for American Elm 1.

6 Scroll through the symbols until you see the 3D trees, choose American Elm 1, and click OK.

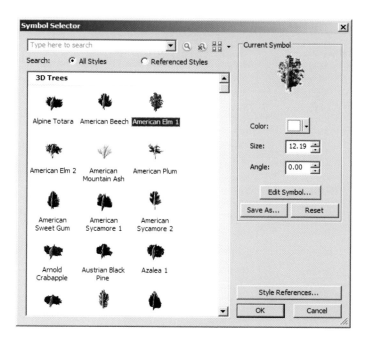

10-1
10-2
10-3
10-4
10-5
10-6
10-7
10-8
10-9
10-10
A10-1
A10-2
A10-3

7 Double-click the symbols for the remaining trees and assign the corresponding 3D symbol. 3D symbols can be assigned for all trees except "stumps."

8 Click OK to close the Layer Properties window.

9 Click the Navigate > Zoom > Pan buttons to view the trees from street level. The image shows trees along a few streets in Pittsburgh's North Shore.

10 Save your ArcScene.

YOUR TURN

Add the point feature class NorthShoreVehicles from 3DAnalyst.gdb. Set the base height for the surface to chapter10_tin, offset the points 5 units, zoom to the layer, and then to a few vehicles. Display the vehicles as 3D vehicle symbols using Unique Values based on the Type value field from the feature attribute table. In the Symbol Selector window, rotate each vehicle 115 degrees by typing a new Angle. Explore the other 3D symbols that come standard with ArcEditor, including 3D Basic, 3D Billboards, 3D Buildings, 3D Industrial, 3D Residential, and 3D Street Furniture. Save the scene when you are finished.

10-1
10-2
10-3
10-4
10-5
10-6
10-7
10-8
10-9
10-10
A10-1
A10-2
A10-3

Tutorial 10-8

Editing 3D objects

You can edit 3D objects using the 3D Editor toolbar. Edits include changing 3D heights, moving 3D objects, or creating new features. In this exercise, you export the proposed North Shore buildings as a new feature class to edit.

Export buildings

1 Save your scene as **Tutorial10-8.sxd** to the Chapter10 folder and zoom to the NorthShoreBldgs layer.

2 In the table of contents, right-click the ProposedBldgs layer and click Data > Export.

3 Click "Use the same coordinate system as the data frame" and save this as a feature class, **\Chapter10.gdb\ProposedBldgsEdits**.

4 Add the new layer to the scene, drape it to the TIN, extrude using [STORIES *10], and change the color to Midday Yellow.

5 Turn the original ProposedBldgs layer off.

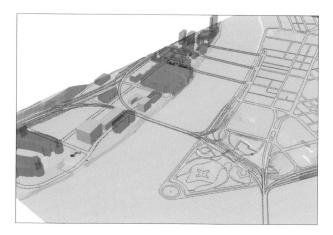

Use the 3D Editor toolbar

1 Click Customize > Toolbars > 3D Editor.

2 On the 3D Editor toolbar, click 3D Editor > Start Editing > ProposedBldgsEdits > OK.

Edit 3D building height

1 On the 3D Editor toolbar, click the Edit Vertex tool.

2 Click one of the ProposedBldgsEdits polygons to select it.

3 On the 3D Editor toolbar, click the Attributes button.

4 In the Attributes window, type **40** for the number of stories and press ENTER.

ArcScene changes the building to reflect the new height.

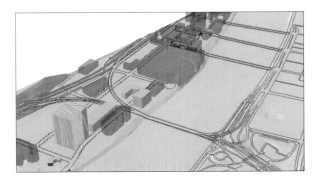

Move a 3D building

1 On the 3D Editor toolbar, click the Edit Placement tool.

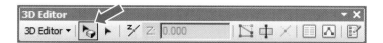

2 Drag a building to a new location on the map. The building is in a new location and drapes to the contours that it was moved to.

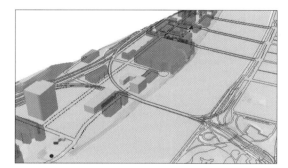

10-1

10-2

10-3

10-4

10-5

10-6

10-7

10-8

10-9

10-10

A10-1

A10-2

A10-3

> ### *YOUR TURN*
>
> Practice moving and editing other proposed building heights. When finished, click 3D Editor, Stop Editing, and Save Edits. Save your scene and exit ArcScene.

Tutorial 10-9

Using ArcGIS 3D Analyst for landform analysis

In this tutorial you explore how to create steepest paths, contour profiles, and line-of-sight analysis using a TIN in ArcMap.

Start a map document

1 Start ArcMap and open a new empty map.

2 Save the map document as **Tutorial10-9.mxd** to the Chapter10 folder.

3 If it is not already open, click Customize > Toolbars > 3DAnalyst to load the toolbar.

4 Click the Add Data button, navigate to the Chapter10 folder, click chapter10_tin > Add. The TIN that you created in Tutorial10-2 is added.

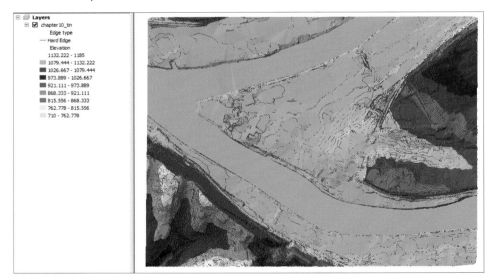

Create a steepest path line and elevation profile

The Create Steepest Path tool calculates the direction a ball would take if released at a given point on the surface. The tool works with raster, triangulated irregular network (TIN), and other surfaces. The result is a 3D graphic line added to the map or scene. You can then create a profile graph of this 3D graphic line.

10-1
10-2
10-3
10-4
10-5
10-6
10-7
10-8
10-9
10-10
A10-1
A10-2
A10-3

1 On the 3D Analyst toolbar, click the Create Steepest Path button.

2 Click a location from the South Shore to Mount Washington as seen in the image.

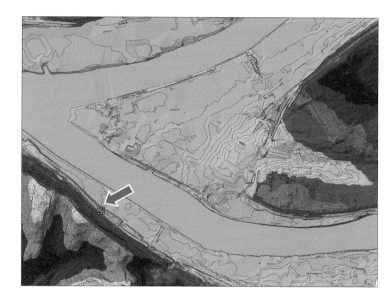

3 On the Tools toolbar, click the Select Elements button, click the lower left grip box, and drag to create a longer path line as shown.

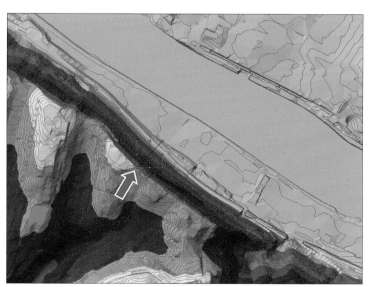

4 On the 3D Analyst toolbar, click the Profile Graph button.

5 Double-click the profile graph title and in the resulting window click the Appearance tab.

6 Type **Topography Analysis** as the Graph profile title, **Change in Elevation from South Shore to Mt Washington** as the footer, and click OK. The graph now reflects the profile analysis.

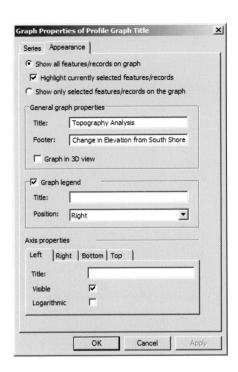

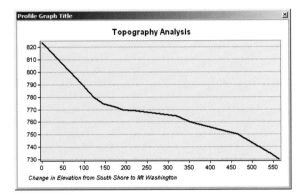

7 Close the profile graph window.

Create a line of sight

A line-of-sight analysis creates a graphic line between two points showing where the view is obstructed between those points.

1 On the 3D Analyst toolbar, click the Create Line of Sight button.

2 In the Line Of Sight window, type **6** as the observer and target offsets. The offset is the number of feet above the TIN for the observer and target. If you set the observer and target heights to zero, then typically you will have a view with more obstructions than one with a height greater than one.

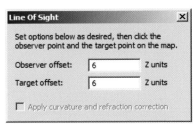

3 Click a point near Pittsburgh's Point where the three rivers meet.

4 Click a location on Mount Washington where the elevation is above 910 feet (gray area on the map). The resulting map shows red along the line where the observer's line of sight is obstructed and green along the line where the view is not obstructed.

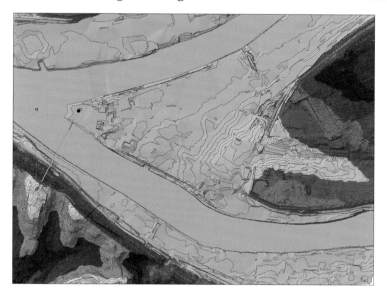

10-1
10-2
10-3
10-4
10-5
10-6
10-7
10-8
10-9
10-10
A10-1
A10-2
A10-3

YOUR TURN

Choose additional observer and target points to see line-of-sight visibility. Change the observer and target heights to see if the visibility changes. Create new steepest line paths and profiles. Explore other 3D Analyst functions in ArcToolbox. Click 3D Analyst Tools, Triangulated Surface to create surface aspect, contours, difference, and slope analysis. Save your map document and exit ArcMap.

Tutorial 10-10

Exploring ArcGlobe

ArcGlobe provides a seamless basemap infrastructure in an ArcScene-like interface for the entire world with imagery, elevation, political boundaries, and highways. You can add your own layers to ArcGlobe and quickly have an impressive GIS application. You need a broadband Internet connection to use the web service that provides the basemaps.

Launch ArcGlobe

1 On your desktop, click Start > All Programs > ArcGIS > ArcGlobe 10 > OK. ArcGlobe opens showing default layers that are provided by Esri.

2 Save the document as **Tutorial10-10.3dd** in your Chapter10 folder. ArcGlobe documents have the extension .3dd.

Explore ArcGlobe

By default, ArcGlobe opened with the Navigate tool selected.

1 Click, hold, and drag the display to the right so that the west coast of North America is at the center of the map.

2 Place your cursor as seen in the image.

3 Right-click and drag downward until San Francisco appears, then re-center the map on San Francisco using your left mouse button. If you don't know where San Francisco is located, you can add the CACounties layer from UnitedStates.gdb, select San Francisco County, and zoom to selected features.

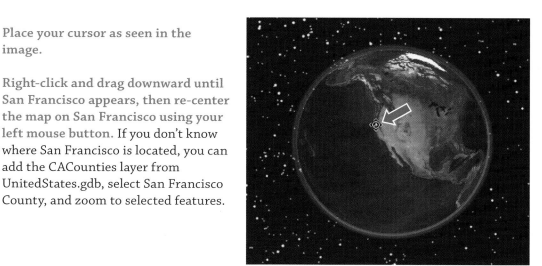

10-1
10-2
10-3
10-4
10-5
10-6
10-7
10-8
10-9
10-10
A10-
A10-
A10-

4 Keep this process up until you can see San Francisco and the Bay Bridge. If you need to zoom out at some point, right-click the map and drag upwards. You can also use your mouse wheel to zoom in and out.

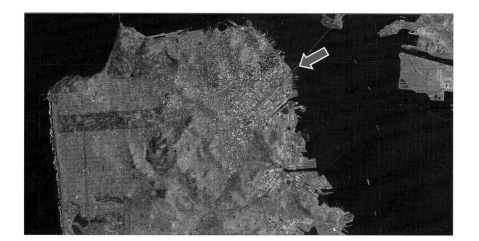

5 Zoom in even closer as indicated in the image. Here you can see the shadow of the Bay Bridge.

6 Click Bookmarks > Create Bookmark and create a bookmark called **San Francisco Bay Bridge**.

Add and display large-scale vector data

You can add and display map layers for anywhere in the world. Next, you add two layers for New York City.

1 Click the Add Data button, browse through the Data folder to NYC.gdb, click Boroughs > Add > Next > Finish > Close.

2 In the table of contents, right-click Boroughs, click Properties > Symbology tab, and change the layer's symbol to no fill and a bright yellow outline with width 1.

3 In the table of contents, right-click Boroughs > Zoom To Layer.

4 In the table of contents, drag the Imagery layer below Boroughs. In the image on the next page, you can see the New York City Boroughs in yellow.

5 Add the layer ManhattanWaterfrontParks from NYC.gdb and symbolize as a green fill, 25% transparent.

6 Zoom to the area of lower Manhattan.

7 Turn the Boroughs layer off. You can see the New York City waterfront parks as polygons and the aerial image.

8 Create a spatial bookmark called **Lower Manhattan**.

YOUR TURN

Practice adding other NYC layers and zoom to various boroughs in New York. Save your ArcGlobe document and close ArcGlobe.

10-1
10-2
10-3
10-4
10-5
10-6
10-7
10-8
10-9
10-10
A10-1
A10-2
A10-3

Assignment 10-1

Develop a 3D presentation for downtown historic sites

Many US cities, including Pittsburgh, are experiencing a surge of downtown revitalization. In Pittsburgh, new condominium and apartment projects are in progress, and the city planning department wants to verify that this new development does not interfere with existing historic sites. In this assignment, you help the city planning department raise the awareness of historic sites in downtown Pittsburgh by developing a 3D model and animation of these areas.

Get set up

- Rename the folder \EsriPress\GIST1\MyAssignments\Chapter10\Assignment10-1YourName\ to your name or student ID. Store all files that you produce for this assignment in this folder.
- Create a new map document called **Assignment10-1YourName.mxd** with relative paths.
- Create a new file geodatabase called **Assignment10-1YourName.gdb**.
- Create a new 3D scene called **Assignment10-1YourName.sxd** with relative paths.

Build the map

Import the following into your file geodatabase:

- \EsriPress\GIST1\Data\Pittsburgh\CBD.gdb\Histsites—polygon features of historic sites in the city of Pittsburgh's Central Business District.
- \EsriPress\GIST1\Data\3DAnalyst.gdb\Bldgs—polygon features of buildings in downtown Pittsburgh.
- \ EsriPress\GIST1\Data\3DAnalyst.gdb\Curbs—line features of curbs (sidewalks) in downtown Pittsburgh.
- \ EsriPress\GIST1\Data\3DAnalyst.gdb\Topo—line features of topography contours in downtown Pittsburgh.
- \ EsriPress\GIST1\Data\3DAnalyst.gdb\Rivers—polygon features of rivers in downtown Pittsburgh.

Process the data

In ArcMap, add Bldgs and Histsites, and create a new feature class of buildings that have their centroids in historic sites and another of buildings whose centroids are not within historic sites. **Hint:** First select buildings in historic sites, and then use Switch Selection in the attribute table to select the nonhistoric site buildings. Save the new features in your file geodatabase **HistSitesBldgs** and **NonHistSitesBldgs**. Remove the original Bldgs layer from the map document.

Requirements

- In ArcScene, add all of the features in your file geodatabase.
- Create a new TIN from the Topo layer called **Assignment10-1YourName_tin** and assign it spatial reference NAD_1983_StatePlane_Pennsylvania_South_FIPS_3702_Feet.
- Remove the original Topo layer and display the TIN as "Faces with the same symbol" (Early Desert color) and no edges or elevation.
- Drape all features to the TIN using a layer offset of 10 for all layers except Rivers. Use a layer offset of 5 for Rivers.
- Extrude the buildings using the stories field multiplied by 10.
- Show the nonhistoric site buildings using a transparency effect of 60% and the historic site buildings as opaque (0% transparency).

Create a PowerPoint presentation

Create a PowerPoint presentation called **Assignment10-1YourName.pptx** that includes the following:

- Title slide with title Downtown Pittsburgh Historic Sites Building Study, including your name and today's date
- Slides with 2D images inserted of the following 3D views:
  - Map extent
  - Map zoomed to historic sites building layer
  - Three additional views of historic sites using target, observer, and various zooms

10-1
10-2
10-3
10-4
10-5
10-6
10-7
10-8
10-9
10-10
A10-1
A10-
A10-

Assignment 10-2

Topographic site analysis

Pittsburgh's Phipps Conservatory (http://phipps.conservatory.org) was built in 1893 by Henry Phipps as a gift to the city of Pittsburgh. The conservatory recently underwent a major renovation with the addition of a 10,885-square-foot green-engineered welcome center, a state-of-the-art production greenhouse, and a one-of-a-kind tropical forest. In 2009, Phipps Conservatory was the site of the welcome dinner for the G20 summit where it hosted an array of world leaders. Near this site, Carnegie Mellon University is expanding and needs to perform line of sight and analysis of the topography. The 3D Analyst extension is very useful for envisioning the area and conducting important site analysis. In this assignment, you use the extension to view the topography of the area, create a 3D TIN, perform a line-of-sight analysis, and conduct slope studies.

Get set up

- Rename the folder \EsriPress\GIST1\MyAssignments\Chapter10\Assignment10-2YourName\ to your name or student ID. Store all files that you produce for this assignment in this folder.
- Create a new map document called **Assignment10-2YourName.mxd** with relative paths.
- Create a new file geodatabase called **Assignment10-2YourName.gdb**.

Build the map

Import the following into your file geodatabase:

- \EsriPress\GIST1\Data\Pittsburgh\Phipps.gdb\Bldgs—polygon features buildings in the Phipps Conservatory study area.
- \EsriPress \GIST1\Data\Pittsburgh\Phipps.gdb\Topo—line features topography contours in the Phipps Conservatory study area.
- \EsriPress \GIST1\Data\Pittsburgh\Phipps.gdb\PhippsAddition—polygon feature of the Phipps Conservatory welcome center addition.

Requirements

- In ArcMap add the Bldgs, Topo, and PhippsAddition. Display Bldgs as a white fill, black outline, and labeled using the NAME field.
- Display the Phipps addition as Mars Red fill with a black outline.
- Create a TIN from the Topo layer called **Assignment10-2YourName_tin** and assign it spatial reference NAD_1983_StatePlane_Pennsylvania_South_FIPS_3702_Feet.
- Create a 36-by-24-inch map layout with three data frames. Include the buildings, Phipps addition, and TIN in all three data frames.
- Rename data 1 **Slope Analysis**, remove the original Topo layer, and display the TIN as "Face slope with a graduated color ramp." Remove Edge Types and Elevation from the symbology and display the slope as a light green to dark green color ramp with 9 quartile classes.

- Rename data frame 2 **Line of Sight Analysis**, display the TIN as "Faces with the same symbol" and a light yellow color. Remove the other Edge Types and Elevation from the symbology. Create a line-of-sight analysis using 6 feet as the observer and target offsets from the Phipps addition to Café Phipps and CMU's Scaife Hall. Create a line-of-sight analysis from Scaife Hall to the Boundary Street building.

- Rename data frame 3 **Topography Analysis**, display the TIN as "Faces with the same symbol" and a light yellow color. Remove the other Edge Types and Elevation from the symbology. Zoom to buildings Scaife Hall and Boundary Street building and create a 3D line from Scaife Hall to the Boundary Street building. This hillside is the location of future CMU building development. Create a 3D profile graph with title **Topography Analysis** and footer subtitle **Change in elevation from Scaife Hall to Boundary Street**. Add the 3D graph to your layout, under data frame 3.

- Add titles, scale bars, a legend for dataframe 1, and other elements you think necessary for your layout and export it as **Assignment10-2YourName.jpg**.

10-1
10-2
10-3
10-4
10-5
10-6
10-7
10-8
10-9
10-1
A10-
A10-
A10-

Assignment 10-3

3D animation of conservatory study area

Get set up

- Rename the folder \EsriPress\GIST1\MyAssignments\Chapter10\Assignment10-3YourName\ to your name or student ID. Store all files that you produce for this assignment in this folder.
- Create a new map document called **Assignment10-3YourName.mxd** with relative paths.
- Create a new file geodatabase called **Assignment10-3YourName.gdb**.
- Create a new 3D scene called **Assignment10-3YourName.sxd** with relative paths.

Build the 3D GIS

Import the following into your file geodatabase:

- \EsriPress\GIST1\Data\Pittsburgh\Phipps.gdb\Bldgs—polygon features buildings in the Phipps Conservatory study area.
- \EsriPress \GIST1\Data\Pittsburgh\Phipps.gdb\PhippsAddition—polygon features of the Phipps Conservatory welcome center addition.
- \EsriPress \GIST1\Data\Pittsburgh\Phipps.gdb\Curbs—line features sidewalk curbs in the Phipps Conservatory study area.
- \EsriPress \GIST1\Data\Pittsburgh\Phipps.gdb\Topo—line features topography contours in the Phipps Conservatory study area.
- \EsriPress \GIST1\Data\Pittsburgh\Phipps.gdb\Image—USGS ortho image in the Phipps Conservatory study area.

Digitize new features

- In your file geodatabase, create a new point feature class called **PhippsTrees** whose spatial reference system is NAD_1983_StatePlane_Pennsylvania_South_FIPS_3702_Feet. In the PhippsTrees attribute table, create a text field called **Type**.
- In ArcMap add the PhippsTrees features and the raster image from your file geodatabase. Using the ortho image as a background, digitize 15 or so trees in the front of Phipps Conservatory and assign them tree types Big Leaf Maple, Italian Cypress 1, and American Sycamore 2.

Requirements

- In ArcScene add all of the features in your file geodatabase. Create a new TIN from the Topo layer called **Assignment10-3YourName_tin** and assign it spatial reference NAD_1983_StatePlane_Pennsylvania_South_FIPS_3702_Feet.
- Remove the original Topo layer and display the TIN as "Faces with the same symbol" (Early Desert color) and no edges or elevation.

- Drape the Bldgs, PhippsAddition, and PhippsTrees to the TIN using a layer offset of 0. Drape the ortho image and Curbs using a layer offset of 5. Depending on your computer's configuration, it might take some time to drape the ortho image.
- Extrude Bldgs and PhippsAddition polygon features using the height fields in each attribute table. The values in these fields are the actual building heights as opposed to the number of stories.
- Show the PhippsAddition using a transparency effect of 50% and the ortho image using a 25% transparency effect.
- Display the PhippsTrees point features as unique symbols using the corresponding 3D symbols for the three tree types.
- Create an animation of the site that eventually zooms into the new Phipps addition. Save the animation as an ArcScene animation file called **Assignment10-3YourName**.
- If instructed to do so, export the animation as **Assignment10-3YourName.avi**.

10-1
10-2
10-3
10-4
10-5
10-6
10-7
10-8
10-9
10-10
A10-1
A10-2
A10-3

Part III
Analyzing spatial data

ArcGIS Spatial Analyst for Desktop

This chapter is an introduction to ArcGIS Spatial Analyst, an extension of ArcGIS for Desktop. With Spatial Analyst you can use or create raster datasets to display and analyze data that is distributed continuously over space as a surface. In this chapter, after an introduction to raster map layers, you prepare and analyze a demand surface map for the location of heart defibrillators in the city of Pittsburgh based on the number of out-of-hospital cardiac arrests. You also learn how to use ArcGIS Spatial Analyst to perform queries for a site suitability analysis and to create a poverty index surface that combines several census data measures from block and block group polygon layers.

Learning objectives

- *Learn about raster map layers*
- *Create a hillshade map*
- *Make kernel density maps*

- *Extract point estimates from raster maps*
- *Conduct a raster-based site suitability study*
- *Build a raster-based risk index*

Tutorial 11-1

Processing raster map layers

The map document that you will open has map layers that include raster maps from a US Geological Survey website, http://gisdata.usgs.net/website/MRLC/viewer. php, *for digital elevation (NED shaded relief, 1/3 arc second) and land use (NLCD 2006). All raster maps are rectangular in their coordinate systems, but you use Pittsburgh's boundary as a mask so that cells outside its boundary have no color and thus are invisible, and the cells inside have their assigned colors. In addition, you use the NED layer to create a hillshade layer, which provides a 3D appearance of topography illuminated by an artificial sun. Placing the hillshade under the land-use layer and giving the land-use layer some transparency makes an attractive and informative display.*

Examine raster map layer properties

1 Open Tutorial11-1.mxd from the Maps folder. The vector map layer called OHCA (out-of-hospital cardiac arrests) is the number of heart attacks over a five-year period per census block that occurred outside of hospitals where bystander help was possible because of location.

11-1

11-2

11-3

11-4

11-5

11-6

A11-

A11-

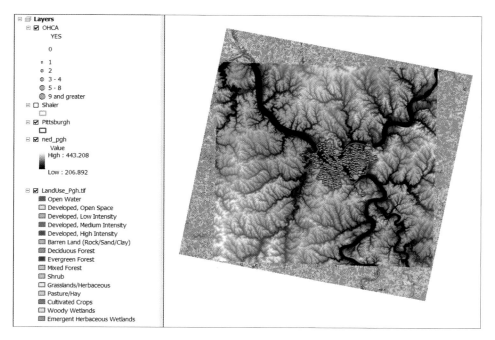

2 Save the map document to the Chapter11 folder of MyExercises.

Next, you examine properties of the raster layers.

3 **In the table of contents, right-click ned_pgh, click Properties > Source tab.** You can see in the Columns and Rows property that this raster map has 2,106 columns and 1,984 rows (and thus over four million cells displayed). You can also see in the Cell Size property that cells are square, 90.7 feet on a side. This layer is not very attractive or informative yet, but it will be after you transform it into a hillshade layer in the next exercise of this chapter.

4 **Scroll down until you see the Statistics properties.** Each cell or pixel has a single value—elevation above mean sea level, feet in this case—which is stored as a floating point number. The statistics for elevation over the extent includes a mean of 323.7 feet, a minimum of 206.9 feet, and maximum of 443.2 feet.

5 Cancel the Properties window.

YOUR TURN

Examine the properties of the LandUse_Pgh layer. Notice in the Spatial Reference properties that this is a projected layer that uses a projection for the continental United States (this is the reason why the layer tilts when ArcGIS reprojects it to the local state plane projection). Also notice in the Raster Information properties that the cell size is about the same as the NED, 30 meters or 98.4 feet on a side, and that the values are integers corresponding to land-use categories.

Set raster environment

Next, you need to set the environment for using the ArcGIS Spatial Analyst tools. Each time you use one of the tools, ArcMap automatically applies the environmental settings, thereby saving you time and effort. First, you have to turn the Spatial Analyst extension on, which is what you do in step 1.

1 Click Customize > Extensions > Spatial Analyst on > Close.

2 On the Menu bar, click Geoprocessing > Environments > Raster Analysis (you may have to scroll down in the Environment Settings window), and type or make selections as shown in the image.

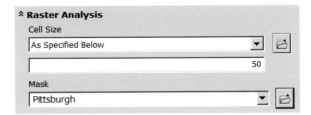

3 Click OK. Now Spatial Analyst automatically uses Pittsburgh as a mask and builds raster layers with cells 50 feet on a side. The unit is feet in this case because the data frame's projection is state plane in feet.

Extract land use using a mask

ArcGIS for Desktop can display a great many raster or image file formats. The land-use layer in your map document is a TIFF (Tagged Image File Format) image as downloaded from the USGS website. To process the layer, it is necessary to convert it to an Esri format, which you do by extracting the rectangular portion of it that is Pittsburgh's extent and saving it in a file geodatabase. At the same time, you use the Pittsburgh boundary as a mask to display cells only within Pittsburgh's boundary.

1 On the Menu bar, click Windows > Search. Type **extract** in the search text box, press ENTER, and click Extract by Mask (Spatial Analyst) to open that tool.

2 Type or make selections as shown in the image. This is one of the few tools that makes you select a mask even though you set the mask in the raster.

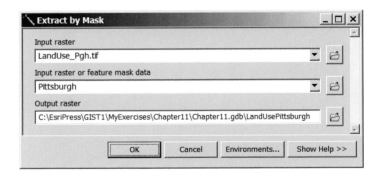

11-1
11-2
11-3
11-4
11-5
11-6
A11-
A11-

3 Click OK. Close the Search window.

4 In the table of contents, turn off all layers except OHCA and LandUsePittsburgh.

5 Right-click LandUsePittsburgh and click Zoom to Layer. ArcMap gave LandUsePittsburgh an arbitrary color ramp, but next you add a layer file to correctly symbolize the new raster map.

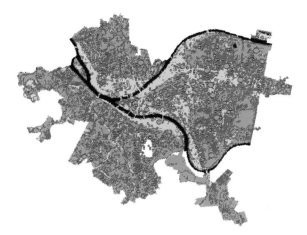

6 Right-click LandUsePittsburgh, click Properties > Symbology tab > Import button ☝ , browse to the Data > SpatialAnalyst folder, click LandUse.lyr > Add > OK > OK. The resulting map is informative and attractive; for example, you can see high-density development along Pittsburgh's rivers, and you can see that the clusters of heart-attack locations are in developed areas. In the next tutorial, you make the map even better by giving it a 3D appearance, using hillshade based on the NED layer.

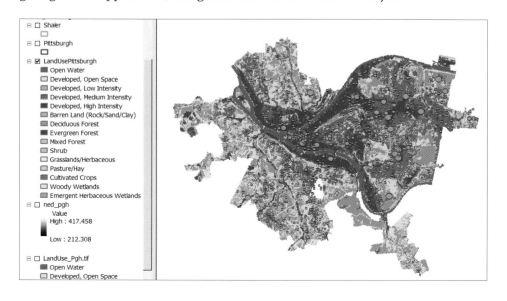

7 Save your map document.

YOUR TURN

Extract LandUseShaler from LandUse_Pgh using the Shaler layer as a mask and saving it as **LandUseShaler** in \EsriPress\GIST1\MyExercises\Chapter11\Chapter11.gdb\. Symbolize the new layer using LandUse.lyr. When finished, turn off all of the Shaler layers and zoom back to the Pittsburgh layer. **Hint:** You have to click Geoprocessing > Environments > Raster Analysis and change the mask to Shaler to do this work. The extract tool does not work unless you make this change, even though you select Shaler as the mask in the tool. After the extraction is finished, be sure to set the Raster Analysis mask back to Pittsburgh. Save your map document and turn all Shaler map layers off in the table of contents.

Tutorial 11-2

Creating a hillshade raster layer

11-1
11-2
11-3
11-4
11-5
11-6
A11-
A11-

The hillshade function simulates illumination of a surface from an artificial light source representing the sun. Two parameters of this function are the altitude of the light source above the surface's horizon in degrees and its angle (azimuth) relative to true north. The effect of hillshade to a surface, such as elevation above sea level, is striking, giving a 3D appearance due to light and shadow. You can enhance the display of another raster layer, such as land use, by making land use partially transparent and placing hillshade beneath it.

Create hillshade for elevation

You will use the default values of the Hillshade tool for azimuth and altitude. The sun for your map will be in the west (315°) at an elevation of 45° above the horizon.

1 Save your map document as Tutorial11-2.mxd.

2 Type **hillshade** in the search text box, press ENTER, click Hillshade (Spatial Analyst), and close the Search window.

3 Type or make selections as shown in the image.

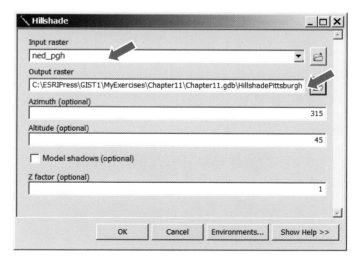

4 Click OK. Notice that Spatial Analyst used the Pittsburgh mask that you set in the raster environment. If your output does not display for Pittsburgh, check Geoprocessing, Environment Settings, Raster Analysis to see if Pittsburgh is selected as the mask, and rerun Hillshade.

5 Right-click HillshadePittsburgh, click Properties > Symbology > Classified (in the Show panel) > Classify button (in the main panel).

6 Select Standard Deviation for Classification Method and 1/3 Std Dev for interval size. Click OK > OK.

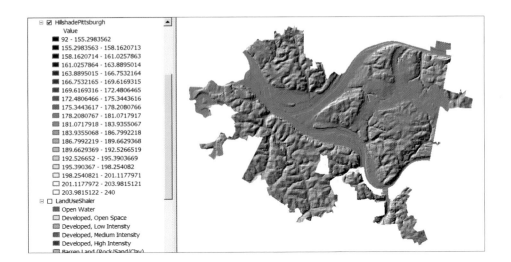

7 Right-click LandUsePittsburgh in the table of contents, and click Properties and the Display tab.

8 Type **35** in the Transparency field and click OK. Turn on LandUsePittsburgh.

9 Move HillshadePittsburgh to just below LandUsePittsburgh in the table of contents. That's the finished product. Hillshade provides a much richer looking map.

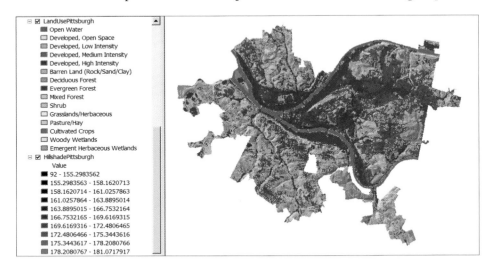

10 Save your map document.

YOUR TURN

Create HillshadeShaler and display it under a 35% transparent LandUseShaler. Be sure to change the mask to Shaler in Geoprocessing > Environments and then back to Pittsburgh.

11-1

11-2

11-3

11-4

11-5

11-6

A11-

A11-

Tutorial 11-3

Making a kernel density map

The incidence of myocardial infarction (heart attacks) outside of hospitals in the United States for ages 35 to 74 is approximately 5.6 per thousand males per year and 4.2 per thousand females per year.[1] You will use a point feature class of census block centroids in Allegheny County as input to an estimation method called kernel density smoothing to analyze heart attack incidence. This method estimates incidence as a density surface (heart attacks per unit area) and has two parameters, cell size and search radius.

Make environmental settings and get statistics

The map document that you open in this tutorial shows actual and estimated values. The observed locations of heart attacks (outside of hospitals and thus with the potential of bystander assistance) are shown with size-graduated point markers at block centroids. Estimated heart-attack incidence at block centroids is shown with a color gradient. The corresponding table of block centroids has the mapped incidence attribute, Inc = 0.0042 × [Fem35T74] + 0.0056 × [Male35T74], where Fem35T74 is the population by block of females ages 34 to 74, and Male35T74 is the population for males. The question is whether estimated incidence computed from residential locations does a good job of estimating the observed heart attacks in the OHCA point file. You will see right away that there are many locations where estimated incidence is high but there are no observed heart attacks, so maybe the estimates aren't very accurate.

1 In ArcMap, open Tutorial11-3.mxd from the Maps folder. See the map on the next page. The vector-based map display for estimated incidence using block centroids with point markers is as good as vector graphics allow. Nevertheless, the map is difficult to interpret; there are too many discrete points.

Next, you create an alternative representation of incidence by estimating a smoothed mean surface using kernel density smoothing. The result is a continuous map that is much easier to interpret and study than the vector map.

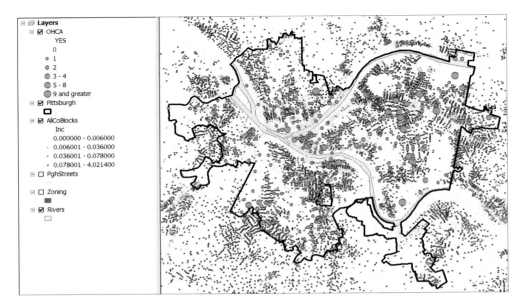

2 Save your map document as **Tutorial11-3.mxd** to the Chapter11 folder.

3 On the Menu bar, click Geoprocessing > Environments > Raster Analysis, and type or make selections as shown in the image on the right.

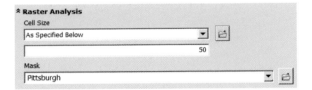

4 On the Menu bar, click Selection > Select By Location > and make selections as shown in the image on the right.

5 Click OK.

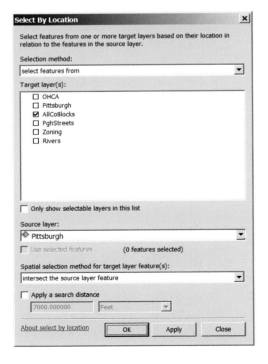

GIS TUTORIAL 1   ArcGIS Spatial Analyst for Desktop **CHAPTER 11** 393

11-1
11-2
11-3
11-4
11-5
11-6
A11-
A11-

6 Right-click AllCoBlocks in the table of contents, click Open Attribute Table, right-click the header for Inc, and click Statistics. Note the sum of Inc, 684 for Pittsburgh, which is the expected annual number of heart attacks in Pittsburgh outside of hospitals in a year's time.

In the next tutorial, you verify that density smoothing preserves this sum in any surface it estimates. Kernel density smoothing simply spreads the total around on a smooth surface, preserving the input total number of heart attacks.

7 Close the Selection Statistics window and the table, and clear the selection.

Make a density map for heart-attack incidence

The OHCA map layer shows heart attacks per census block in Pittsburgh. Blocks in Pittsburgh average a little less than 300 feet per side in length. Suppose that policy analysts estimate that a defibrillator with public access can be made known to residents and retrieved for use as far away as 2.5 blocks from the location. They thus recommend looking at areas that are five blocks by five blocks in size, or 1,500 feet on a side, with defibrillators located in the center. Therefore, you can use a 150-foot cell and a 1,500-foot search radius. The 150-foot cell approximates the middle of a street segment, the average location of a heart attack.

1 Type **kernel density** in the search text box, press ENTER, click Kernel Density (Spatial Analyst), and close the Search window.

2 Type or make selections as shown in the image.

3 Click OK.

The resulting surface will appear useful after you symbolize it better in the following step.

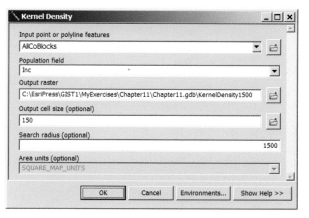

4 Right-click KernelDensity1500, click Properties > the Symbology tab > the Classify button. Select Standard Deviation for Classification Method. In general, Standard Deviation is a good option for showing variation in raster grids because it yields a central category and an equal number of categories on each side of the center. That makes dichromatic color scales more meaningful and easier to interpret. You control the number of categories in the next step by choosing the fraction of standard deviation for creating break points, every 1, 1/2, 1/3, or 1/4 standard deviation.

Next, you use dichromatic color scales.

5 Select 1/3 Std Dev for Interval Size and click OK.

6 Select the color ramp that runs from green to yellow to red and click OK.

7 Turn off AllCoBlocks. Incidence matches clusters of the OHCA heart-attack data in many, but not all areas. For example, there is a cluster in Pittsburgh's Central Business District (triangle just to the right of where the three rivers meet), but estimated incidence is low there. The problem is that the density map, based on population data, shows expected heart attacks per square foot in reference to where people live, not necessarily where they have heart attacks. Many people shop or work in the Central Business District and unfortunately have heart attacks there, but few live there.

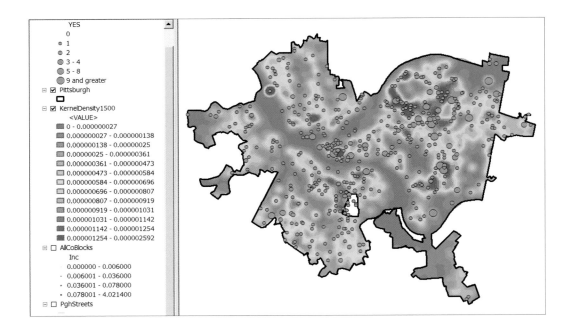

Check the density surface, to see if it preserves the total number of heart attacks. The estimated incidence that you found using block centroids was 684. Open the properties for the density surface and click the Source tab. There you will find the mean of 0.000000420184 heart attacks per square foot. To get the number of cells in the Pittsburgh mask, you have to use a "trick." Select Symbology in the Layer Properties window and click Classify. There you will find that there are 72,315 cells in the Pittsburgh mask. (The estimate of the mean density in this display, 0.000000417, is less accurate than the one from the Source tab.) Remember that each cell is 150 feet by 150 feet. Therefore, 150 × 150 × 72,315 × 0.000000420184 = 684 smoothed heart attacks. So kernel density smoothing

11-1

11-2

11-3

11-4

11-5

11-6

A11-

A11-

moved the input number of heart attacks located at block centroids around to distribute them smoothly. The kernel density map is a better estimate of incidence than raw data, because smoothing averages out spatial randomness and provides an estimate of the mean or average surface. Close the Classification and Layer Properties windows.

YOUR TURN

Create a second kernel density surface for incidence, called KernelDensity3000, with all inputs the same except you will use a search radius of 3,000 instead of 1,500. Symbolize the output the same as Density1500. While keeping KernelDensity1500 turned on, turn KernelDensity3000 on and off to see the differences in the two layers. KernelDensity3000 is more spread out and smoother, but it has the same corresponding number of estimated heart attacks: close to 684. Save your map document.

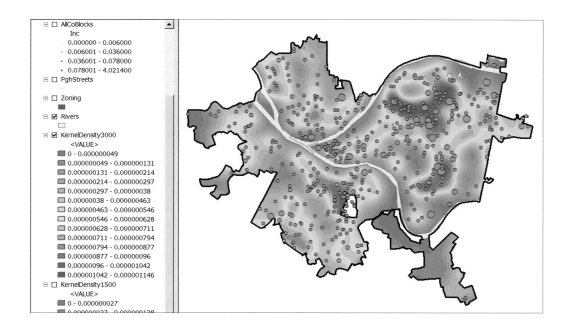

Tutorial 11-4

Extracting raster value points

While the estimated densities appear to match the actual heart-attack data in OHCA, the match may or may not stand up to closer investigation. ArcMap has a tool that extracts point estimates from the raster surface for each point in OHCA. Then you can use the extracted densities multiplied by block areas to estimate number of heart attacks. If there is a strong correlation between the estimated and actual heart attacks, there would be evidence that population alone is a good predictor of heart attacks outside of hospitals.

1 Save your map document as **Tutorial11-4.mxd** to the Chapter11 folder.

2 Type **extract values to points** in the search text box, press ENTER, click Extract Values to Points (Spatial Analyst), and close the Search window.

3 Type or make selections as shown in the image.

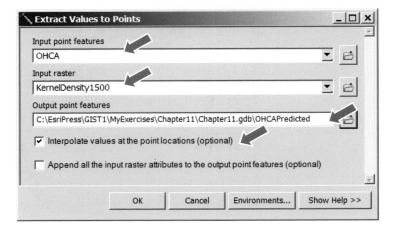

4 Click OK. The resulting layer, OHCAPredicted, has an attribute, RASTERVALU, which is an estimate of heart-attack density, or heart attacks per square foot, in the vicinity of each block.

Calculate predicted heart attacks

1 Right-click OHCAPredicted and open its attribute table.

2 Click Options, Add Field, and add a field called **Predicted** with the Float data type.

11-1
11-2
11-3
11-4
11-5
11-6
A11-
A11-

3 Right-click the Predicted header and click Field Calculator > Yes.

4 Create the expression **5 * [RASTERVALU] * [Area]** and click OK. OHCA data is a five-year sample for heart attacks, so the expression includes the multiple 5.

5 Close the attribute table. A few of the points in OHCA have no raster values near them, so ArcGIS assigns -9999 to their RASTERVALU cells to signify missing values. Before looking at a scatterplot of predicted and actual values, you first select only OHCA points with positive predicted values.

6 Click Selection, Select By Attributes.

7 For the OHCAPredicted layer, create the expression **"Predicted" >= 0** and click OK.

8 Right-click OHCAPredicted, and click Data > Export Data.

9 Export selected features to **\EsriPress\GIST1\MyExercises\Chapter11\Chapter11. gdb\OHCAPredicted2** and click Yes to add the feature class to the map.

10 Clear the selected features and turn off the OHCAPredicted layer.

Create a scatterplot of actual versus predicted heart attacks

While Pittsburgh has a total of 7,466 blocks, only 1,509 blocks had heart attacks. The scatterplot that you will eventually construct includes data for only the 1,509 blocks, but should include the balance of the total blocks, which had actual values of zero but predicted values sometimes much larger than zero. Nevertheless, you can get an indication of the correlation between predicted and actual heart attacks. Adding the balance of blocks would only make the correlation worse, but the correlation is actually already very low, as you see next.

1 Click View > Graphs > Create Graph.

2 Type or make selections as shown on the next page.

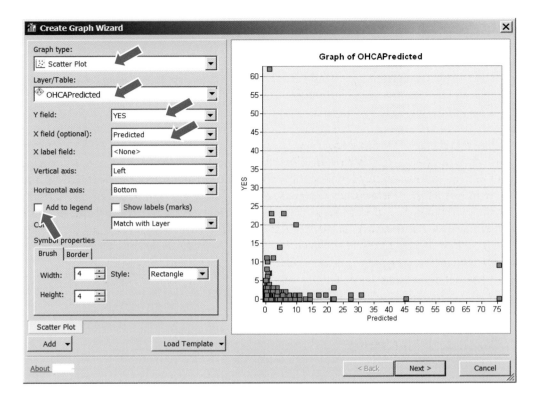

3 Click Next > Finish. At the scale of blocks, the predicted values correlate poorly with the actual values. A good correlation would have a graph with Actual (YES attribute) and predicted values scattering around a 45-degree slope line. This scatterplot shows no correlation at all. If you export the corresponding data to a statistical package or Excel, you would find that the correlation coefficient between predicted and actual values is only 0.0899, which is very low. Evidently, factors other than where the population resides affect the locations and clustering of heart attacks occurring outside of hospitals.

4 Close the graph and save your map document.

11-1
11-2
11-3
11-4
11-5
11-6
A11-1
A11-2

Tutorial 11-5

Conducting a raster-based site suitability study

The objective is to find locations that have high heart-attack rates and that are candidates to have heart defibrillators accessible to the public. The approach starts by using kernel density smoothing on the available heart-attack (OHCA) data to remove spatial randomness from the spatial distribution and provide a more reliable estimate of demand. An assumption is that any locations within commercial areas provide public accessibility. Efforts then turn to querying the kernel density map of heart attacks for areas with sufficiently high densities that are in commercial areas. To query raster maps, you have to reclassify cells to 1 and 0 values (representing true and false) and use Boolean operators, such as AND or OR, to combine raster maps and meet query criteria.

Open a map document

A vector map layer is available for commercial area boundaries. To conduct a raster-based analysis, you have to convert this map layer into a raster layer.

1 **In ArcMap, open Tutorial11-5.mxd from the Maps folder.** The map document shows the observed locations of heart attacks outside of hospitals, a 600-foot buffer of commercially zoned areas in Pittsburgh, and other supporting layers. The 600-foot (or two-block) buffer of commercial areas includes adjacent noncommercial areas that have sufficient access to defibrillators.

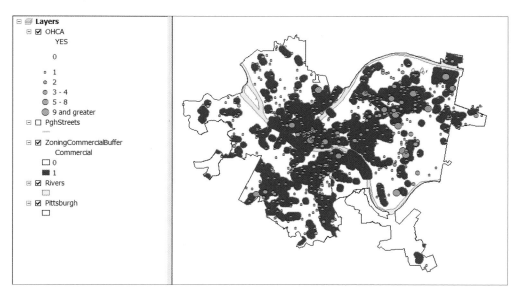

2 Save the map document to the Chapter11 MyExercises folder.

3 On the Menu bar, click Geoprocessing > Environments > Raster Analysis and type or make selections as shown in the image.

4 Click OK.

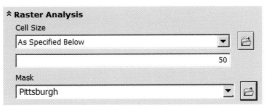

Convert a feature buffer layer to a raster dataset

The ZoningCommercialBuffer layer has two polygons and corresponding records with a single attribute: Commercial. The Commercial value of 1 corresponds to commercial land use or land within 600 feet of commercial land use. The other value, 0, represents the balance of Pittsburgh and includes all other zoned land uses. You will convert this vector layer into a raster dataset using a conversion tool. First, however, you need to select both records in the vector file in order for them to convert.

1 Right-click the ZoningCommercialBuffer layer and click Open Attribute table.

2 Select both records by clicking the row selector of the first row and dragging the mouse to select both rows, and then close the table.

3 Type **polygon to raster** in the search text box, press ENTER, click Polygon to Raster (Conversion), and close the Search window.

4 Type or make selections as shown in the image.

5 Click OK.

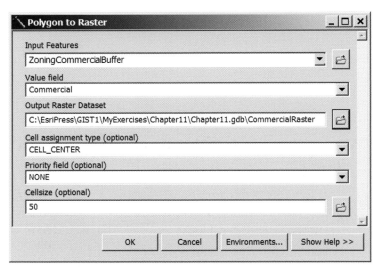

6 Remove the ZoningCommercialBuffer layer and turn off the OHCA layer.

GIS TUTORIAL 1   ArcGIS Spatial Analyst for Desktop   CHAPTER 11   401

11-1
11-2
11-3
11-4
11-5
11-6
A11-1
A11-2

7 Right-click CommercialRaster, click Properties > the Symbology tab > Unique Values in the Show panel, and resymbolize the new Commercial area to have two colors: white for noncommercial and gray for commercial.

YOUR TURN

Create a kernel density map based on the YES attribute of the OHCA point layer that has 150-foot cells, a search radius of 1,500 feet, and area units of SQUARE_FEET. Call the new raster layer **HeartAttack** and save it to Chapter11.gdb. Symbolize the layer using the standard deviation method with interval size 1/3 Std Dev. Use the green-to-yellow-to-red color ramp. Try turning the OHCA layer on and off to see how well the density surface represents heart attacks, and then remove the OHCA layer.

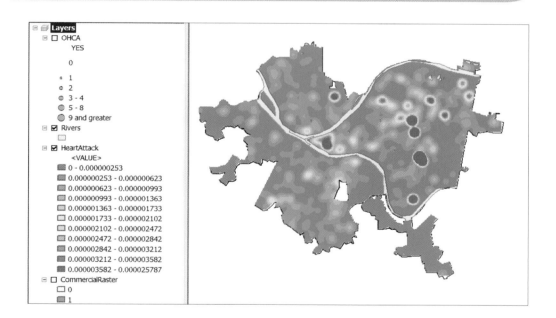

Query a raster dataset with a single criterion using reclassify

First, you reclassify your kernel density map, HeartAttack, for areas that have sufficiently high heart-attack density to merit a defibrillator. Suppose that policy makers seek 25-block areas, roughly five blocks on a side, that would have 10 or more heart attacks every five years in locations where bystander help is possible. A square 25-block area is 5×300 feet = 1,500 feet on a side with 1,500 feet $\times$ 1,500 feet = 2.25×10^6 square feet of area.

Thus, the heart-attack density sought is 10 heart attacks / 2.25×10^6 square feet = 0.000004444 heart attacks per square foot or higher. While the density map you just created has a continuous range of values, next you reclassify values into just two values: 0 for cells with density less than 0.000004444, and 1 for cells with density greater than or equal to 0.000004444.

1 Type **reclassify** in the search text box, press ENTER, click Reclassify (Spatial Analyst), and close the Search window.

2 Select HeartAttack for the input raster and click Classify.

3 Select 2 for Classes, select Manual for Method, type **0.000004444** to replace 0.00003 (the first of two values) in the Break Values panel, and click OK. The Old values column shows values with only 6 decimal places, but the Reclassify tool has all 9 decimal places saved.

4 Click Precision, select 8 for the number of decimal places, and click OK.

5 Finish filling in the form by typing or making selections as shown in the image.

6 Click OK.

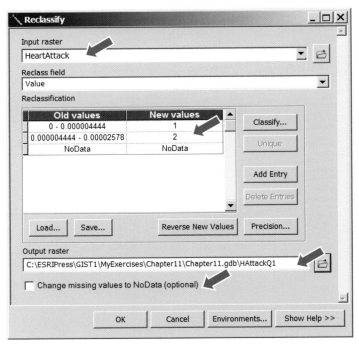

7 Resymbolize HAttackQ1 so that 0 has no color and 1 is dark blue, and make sure that HeartAttack is turned on and below HAttackQ1. In the image on the next page, you can see that relatively few peak areas, six, have sufficiently high heart-attack density and may be large enough in area.

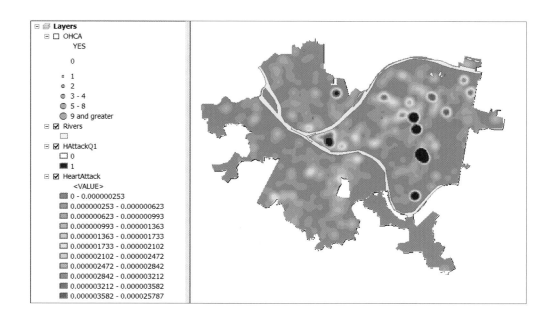

11-1
11-2
11-3
11-4
11-5
11-6
A11-
A11-

Query a raster dataset with two criteria

Next, you include a second criterion in the query—locations within the commercial buffer—for suitable defibrillator sites. Recall that the CommercialRaster map has 0 for noncommercial areas and 1 for commercial areas. In the previous exercise you just set sufficiently dense areas for heart attacks to 1 and those not dense enough to 0. The Boolean AND tool combines two raster datasets by giving all cells the value 0 except where both input cells are 1, in which case the output cell gets the value 1. In this case, the resulting areas defined by cells with value 1 are both in the commercial buffer and have the sufficiently high heart-attack density of HAttackQ1.

1 Type **boolean and** in the search text box. Notice that all of the Boolean logical operators are available as tools, including NOT, AND, and OR.

2 Press ENTER, click Boolean And (Spatial Analyst), and close the Search window.

3 Type or make selections as shown in the image.

4 Click OK.

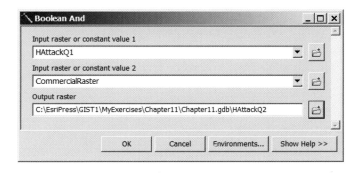

5 Resymbolize HAttackQ2 so that 0 has no color and 1 is Tourmaline Green (column 8, row 3 of the color chips array), and make sure that HAttackQ1 and HeartAttack are turned on and below HAttackQ2. As you would expect, adding a second criterion with the AND connection has reduced the size of areas meeting criteria. Three of the formerly promising areas are significantly reduced.

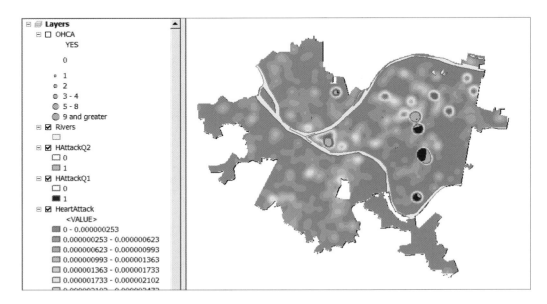

6 Save your map document.

YOUR TURN

Turn on the Streets layer and zoom in to each feasible area to check the third criterion that there be at least 25 blocks, or roughly 2.25 million square feet, in a square area. Use the Measure tool on the toolbar to measure feasible areas. Which areas remain feasible? What would you report back to policy makers? Save the map document.

11-1
11-2
11-3
11-4
11-5
11-6
A11-1
A11-2

Tutorial 11-6

Using ModelBuilder for a risk index

People who live in poverty often have poor health care, unhealthy diets, and unhealthy habits such as smoking—all factors contributing to heart attacks. In this tutorial, you create an index for identifying poverty areas by combining four poverty indicators:[2] (1) population below the poverty income line, (2) female-headed households with children, (3) population 25 or older with less than a high school education, and (4) workforce males 16 or older who are unemployed.

Robyn Dawes provides a simple method for combining such measures into a poverty index.[3] If you have a reasonably good theory that several variables are predictive of a dependent variable of interest (and whether the dependent variable is observable or not), then Dawes makes a good case that all you need to do is to remove scale from each input, so each has the same weight, and then average the scaled inputs to create a predictive index. A good way to remove scale from a variable is to calculate z-scores, subtracting the mean and then dividing by the standard deviation for each variable.

You can see in the following table that if you simply averaged the four variables, then "Male unemployed" arbitrarily gets the highest weight while "Female-headed households with children" would have practically no weight, given the means of the variables. The z-scores for all four variables, however, all have means of zero and standard deviations of one, so when averaged they each will have equal weight.

Indicator variable	Mean	Standard Deviation
Female-headed households with children	1.422.	4.431
Less than high school education	110.060	80.812
Male unemployed	154.500	124.804
Poverty income	126.021	147.188

There are three steps in the workflow for creating the poverty index. First, you calculate the z-scores for each of the four indicators. The second step is to create kernel density maps for all four input variables. The map layers for the indicators are centroids of blocks for the population of female-headed households with children and centroids of block groups for the other three indicators (which are not available at the more desirable, smaller block level). Thus, to make these layers comparable for combining them into an index, you transform them into kernel density maps, all with the same 50-foot-square grid cells. The third step is to use a Spatial Analyst tool to add the surfaces, weighted by 0.25, to average them. You carry out the second and third steps using ModelBuilder to document the work and provide a reusable tool for creating an index. Note that you can work through the following exercises successfully even if you did not complete the introduction to ModelBuilder in tutorial 6-7. Tutorial 6-7 has a more complete introduction to ModelBuilder.

Set the geoprocessing environment

The map document you open has inputs for preparing the poverty index: AllCoBlkGrps, which has block group centroids and needed attributes (NoHighSch = population with less than high school education, Male16Unem = males in the workforce who are unemployed, and Poverty = population below poverty income), and AllCoBlocks, which has block centroids and the attribute FHHChld = female-headed households with children.

1 **In ArcMap open Tutorial11-6.mxd from the Maps folder and save it to the Chapter11 MyExercises folder.** Shown are the block group centroids and block centroids, each displaying one of the four poverty indicators via a color ramp. You can see that it is difficult to represent the spatial patterns with so many points effectively using vector graphics, plus it is difficult to integrate the information from just two spatial distributions out of the four needed for the poverty index. The raster poverty index that you create does a good job on both issues.

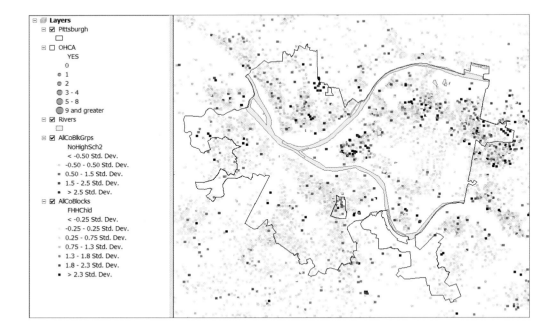

2 **On the Menu bar, click Customize > Extensions > Spatial Analyst to turn on the Spatial Analyst Extension, if necessary.**

3 Click Geoprocessing > Environments > Raster Analysis and make selections or type as shown in the image.

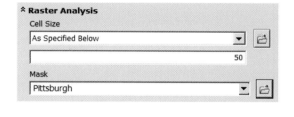

4 Click Geoprocessing > Geoprocessing Options, and make selections as shown in the image.

5 Click OK.

11-1
11-2
11-3
11-4
11-5
11-6
A11-1
A11-2

Standardize input variables

Next you calculate the z-score in an attribute table of one of the input feature classes. To save time, the other three variables already have z-scores ready for use.

1 In the table of contents, right-click AllCoBlocks and click Open Attribute Table.

2 Scroll to the right, right-click FHHChld, and click Statistics. It is convenient to copy and paste the statistics to Notepad and then copy and paste them later to the field calculator that you use.

3 On your desktop, click Start > All Programs > Accessories > Notepad.

4 Select all of the statistics in the Statistics of AllCoBlocks window, press CTRL+C, click inside the Notepad window, and press CTRL+V.

5 Close the Statistics of AllCoBlocks window, click the Table Options button in the Table window > Add Field, type **ZFHHChld** for Name, select Float for Type, and click OK.

6 Right-click the header for ZFHHChld, click Field Calculator > Yes and create the following expression in the bottom panel of the Field Calculator window by copying and pasting from your Notepad window where needed: ([FHHChld] – 1.422147) / 4.431302. Recall that to calculate z-scores for data in a sample, you subtract the mean from each data point and divide by the standard deviation, exactly what you are doing here.

7 Click OK. Below are the first six values for the calculated z-scores.

FHHChld	ZFHHChld
0	-0.320932
0	-0.320932
0	-0.320932
2	0.130403
4	0.581737
1	-0.095265

8 Close the table and close Notepad without saving.

Create a new toolbox and model

1 Click Windows > Catalog, and expand Home–Chapter 11 in the folder/file tree.

2 Right-click Home–Chapter 11, click New and Toolbox, and rename the new toolbox **Index.tbx**.

3 Right-click Index.tbx and click New > Model.

4 In the Model window, click Model > Model Properties.

5 On the General tab, for Name type **PovertyIndex** (no spaces allowed), for Label type **Poverty Index**. Click OK, hide the Catalog window, and leave your Poverty Index window open.

Use kernel density tools for inputs

The next task is to create kernel density layers for the four inputs using the z-scores. After you set up one kernel density tool, you can easily copy and paste it and make adjustments for the remaining three.

GIS TUTORIAL 1   ArcGIS Spatial Analyst for Desktop   CHAPTER 11   409

11-1
11-2
11-3
11-4
11-5
11-6
A11-
A11-

1   Search for **kernel density**, drag Kernel Density (Spatial Analyst) to the Poverty Index model window, and drop it in.

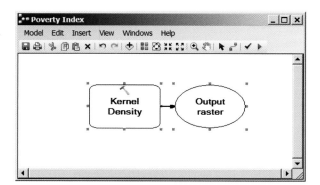

2   In the model, right-click Kernel Density and click Open.

3   Type or make the selections as shown in the image.

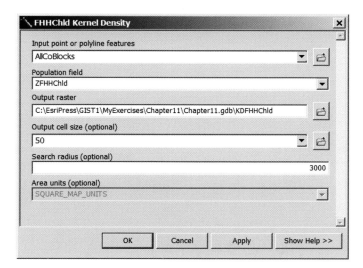

4   Click OK. That adds AllCoBlocks as the input to the kernel density tool. When diagram elements have color fill, as they now do, that portion of the total model is ready to run. While building a model it's a good idea to run parts as you create them, but after the model is finished you generally run the entire model at once.

5   Right-click the Kernel Density tool element, click Rename, and change the name to **FHHChld Kernel Density**.

6 Right-click FHHChld Kernel Density and click Run. Close the Run window when it finishes.

7 Right-click KDFHHChld in the model and click Add to Display.

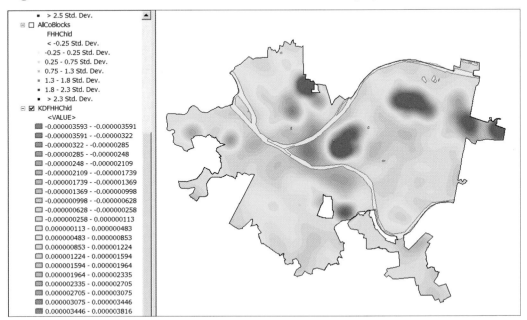

YOUR TURN

Symbolize the new layer using the Classified method with 1/4 standard deviations and the color ramp that runs from blue to yellow to red (11th ramp from the top of those available). Turn off the point feature layers.

Create a kernel density layer for a second input

You can reuse the model elements you just built.

1 In the Model window, right-click FHHChld Kernel Density and click Copy.

2 Click Edit > Paste.

3 Right-click the new FHHChld Kernel Density 2 model element and rename it **NoHighSch Kernel Density**.

11-1

11-2

11-3

11-4

11-5

11-6

A11-1

A11-2

4   Right-click NoHighSch Kernel Density and click Open. Ignore the error message.

5   Type or make the selections as shown in the image.

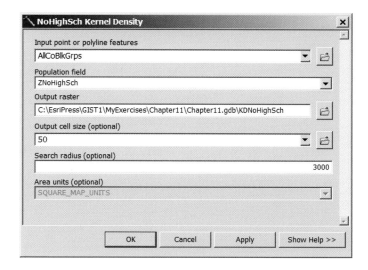

6   Click OK.

7   Right-click KDFHHChld(2) and rename it **KDNoHighSch**.

8   Right-click NoHighSch Kernel Density, click Run, and resymbolize as you like but with a different color ramp from KDFHHChld.

YOUR TURN

Copy and paste the NoHighSch Kernel Density model element two times to use block group attributes ZMaleUnem and ZPoverty to create two new raster layers. See the resulting partial model at right for element names that you need to use. Then run each of the two new model elements and resymbolize resulting map layers. Examine each of the four raster maps. See that they have overlapping but different patterns. The index combines these patterns into a single, overall pattern. Resize and rearrange model elements.

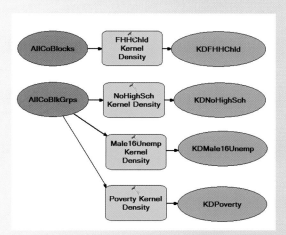

Notice that the process model elements acquire drop shadows in the model window after you run them. To reset the model so that you can run it again, click Model > Validate Entire Model. ModelBuilder removes the drop shadows. Save your model.

Average kernel density maps

1 Type **weighted sum** in the search text box and press ENTER.

2 Drag the Weighted Sum (Spatial Analyst) to your model, to the right of the kernel density outputs, drop it in, and close the Search window.

3 Right-click Weighted Sum, click Open, and type or make selections as shown in the image, including changing weights from 1 to 0.25.

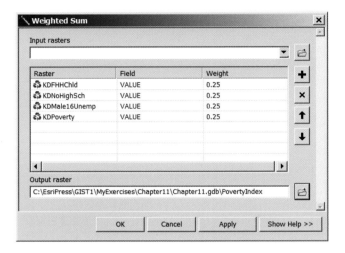

4 Click OK.

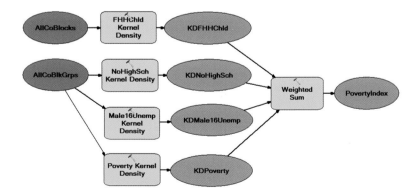

5 Run the Weighted Sum process, right-click PovertyIndex and click Add To Display. Resymbolize the resulting PovertyIndex using Classified in the Show panel, standard deviations, 1/4 Std Dev interval size, and the green-to-yellow-to-red color ramp.

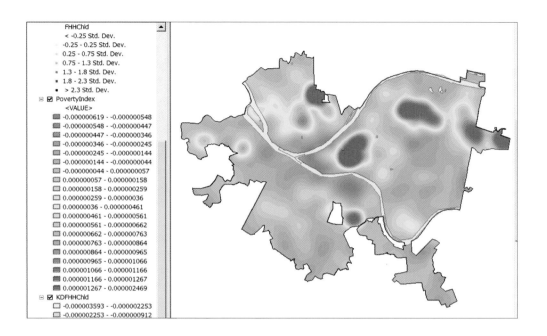

11-1
11-2
11-3
11-4
11-5
11-6
A11-
A11-

6   Save your model.

YOUR TURN

Suppose that you do not care to see the individual kernel density maps, but only the poverty index. Right-click each output of the kernel density tool and deselect Add To Display. Besides preventing future display of these maps, these actions remove the current maps from the map document. Now, right-click PovertyIndex in the table of contents and save a layer file. Right-click PovertyIndex in the model, click Properties > Layer Symbology and select the layer file. Remove PovertyIndex from the table of contents and use Catalog to delete it from Chapter11.gdb. Click Model > Validate Entire Model and then Model > Run Entire Model. After the model runs, you should see the end result, the PovertyIndex raster map layer, in your map document with symbology applied from the PovertyIndex.lyr layer file. Save and close your model but leave the map document open.

Create poverty contour

Suppose that after consideration, policy analysts wish to use the poverty index density of 0.0000009 or higher to define poverty. Next, you create a feature class that has the contour line for that index "elevation."

1 In the search box, type **contour list**, click Contour List (Spatial Analyst), and close the Search window.

2 Type or make the selections as shown.

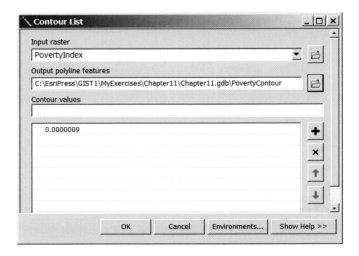

3 Click OK.

4 In the table of contents turn off PovertyIndex. You now have polygons that explicitly define poverty areas and can be used for many policy purposes.

5 Save the map document and close ArcMap.

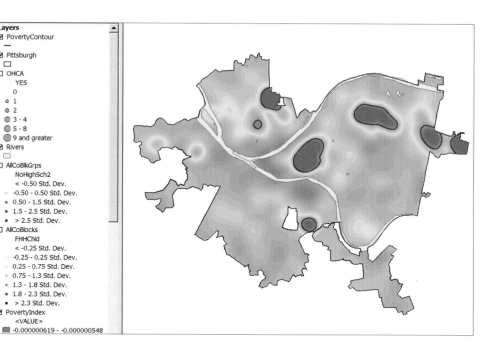

11-1

11-2

11-3

11-4

11-5

11-6

A11-

A11-

Assignment 11-1

Create a mask and hillshade for suburbs

The first ring of suburbs around urban areas is a good area for revitalization as aging suburban homeowners attempt to downsize houses and relocate closer to work. The houses in these areas tend to be relatively small and well-constructed, but in need of renovation.

This assignment has you choose a subset of municipalities in Allegheny County comprising the first ring of suburbs around Pittsburgh to display land use with hillshade. You also display parks for the suburbs. The resulting map document and layers provide a good starting point for redevelopment work.

Get set up

- Create a new folder, \EsriPress\GIST1\MyAssignments\Chapter11\Assignment11-1YourName\ and store all files that you produce for this assignment in it.
- Create a map document called **Assignment11-1YourName.mxd** with relative paths.
- Create a new file geodatabase called **Assignment11-1YourName.gdb** and store finished files that you create in this assignment in it. Do not include or hand in any of the raw data shapefiles in the next bulleted list.

Build the map

Add the following to your map document. Add the municipalities first, so that your data frame inherits that layer's projection, which is the local 1983 State Plane projection.

- \EsriPress\GIST1\Data\AlleghenyCounty.gdb\Munic—polygon features for municipalities in Allegheny County.
- \EsriPress\GIST1\Data\AlleghenyCounty.gdb\Parks—polygon features for parks. DESCR is an attribute with park name.
- \EsriPress\GIST1\Data\SpatialAnalyst\SpatialAnalyst.gdb\NED—digital elevation model for a portion of Allegheny County.
- \EsriPress\GIST1\Data\SpatialAnalyst\SpatialAnalyst.gdb\Pittsburgh—polygon feature for Pittsburgh's municipal boundary.
- \EsriPress\GIST1\Data\SpatialAnalyst\LandUse\LandUse_Pgh.tif—land use for a portion of Allegheny County.

Requirements

- Turn off all layers except Munic to simplify the next step, in which you create a ring of suburbs. Define the first ring of suburbs as those within a distance of 1.25 miles of Pittsburgh, but not including Pittsburgh. Start by making Munic the only selectable layer. Then use Selection, Select By Location, and select municipalities that are within a distance of 1.25 miles of Pittsburgh. Use the Select Features tool, hold down your SHIFT key, and click inside the Pittsburgh polygon in the Munic layer to deselect it. You won't see any

11-1

11-2

11-3

11-4

11-5

11-6

A11-1

A11-2

change in the selection because surrounding municipalities share Pittsburgh's boundary, but the Pittsburgh polygon should no longer be selected. Also deselect Mount Oliver, which is the polygon inside of Pittsburgh and not a suburb. Finally, right-click Munic, click Data > Export data to create **Suburbs**, and add it to your map document. Now select Parks that intersect with suburbs and create **SuburbanParks**.

- Using Suburbs as the mask and cell size of 50, extract a raster from LandUse called **LandUseSuburbs**. Create a hillshade from NED called **HillshadeSuburbs**.

The following are additional instructions on symbolization:

- To symbolize LandUseSuburbs, click Unique Values in the Show panel of the Layer Properties Symbology tab, click the Import button (file icon), browse to the Data > SpatialAnalyst folder, and import LandUse.lyr.

- Use hollow fill and a thin outline for Suburban Parks. Label them with their names with the smallest font size when zoomed in to 1:100,000 or closer (see the Labels tab of the properties sheet and its Scale Range button).

- Symbolize Suburbs with a hollow fill and thick outline. Label with a small font and a halo.

- Symbolize HillshadeSuburbs using Classified in the Symbology show panel with the Standard Deviation classification method and 1/3 Std Dev interval size.

- Make LandUseSuburbs transparent, move it above the hillshade, turn off unneeded layers, and display suburban parks with the shaded land-use layer. Housing will be in the red, developed areas.

Assignment 11-2

Estimate heart-attack fatalities outside of hospitals by gender

Unfortunately, females have more fatal heart attacks outside of hospitals than males, perhaps because symptoms of heart attacks in females are less well known and more subtle than those for males. Heart attacks outside of hospitals are roughly 1.5 per thousand for males age 35 to 74 and 2.3 per thousand for females in the same age range. In this assignment you create two density map layers—one for males and one for females—using these incidence rates for the municipality of Wilkinsburg in Allegheny County. You do all raster processing using Spatial Analysis tools in a model.

Get set up

- Create a new folder, \EsriPress\GIST1\MyAssignments\Chapter11\Assignment11-2YourName\ and store all files that you produce for this assignment in it.
- Create a map document called **Assignment11-2YourName.mxd** with relative paths.
- Create a new file geodatabase called **Assignment11-2YourName.gdb** and store files that you create in this assignment in it. Do not include or hand in any of the raw data shapefiles in the next bulleted list.

Build the map

Add the following to your map document:

- \EsriPress\GIST1\Data\AlleghenyCounty.gdb\Munic—polygon features for municipalities in Allegheny County.
- \EsriPress\GIST1\Data\SpatialAnalyst\SpatialAnalyst.gdb\AllCoBlocks—point features for census block centroids in Allegheny County.

Requirements

- Select the Wilkinsburg polygon from Munic and export it as **Wilkinsburg**. Extract the Wilkinsburg blocks from AllCoBlocks and save them as **WilkinsburgBlocks**. Add floating point fields to the attribute table for WilkinsburgBlocks: **MMortinc = 0.0015 × [Male35T74]** for the annual number of heart-attack fatalities for males age 35 to 74 and **FMortInc = 0.0023 × [Fem35T74]** for females age 35 to 74.
- Create kernel density map layers for MMortinc and FMortinc using Wilkinsburg as the mask and with a cell size of 25 and search radius of 1,500 feet. Add outputs to the map.

Additional instructions

- After you extract Wilkinsburg, to give your data frame Wilkinsburg as its full extent instead of Allegheny County, do the following: Right-click Layers in the table of contents, click Properties, the Data Frame tab. Select Fixed Extent for Extent, click Specify Extent, the Outline of Features option button, Select Wilkinsburg as the layer, and click OK > OK.

- Set the environment for ArcToolbox. For Workspace make \MyAssignments\Chapter11\ Assignment11-2YourName\ the Current Workspace and set the extent to Wilkinsburg. For Raster Analysis, set cell size to 25 and set the Mask to be Wilkinsburg.

- Set Geoprocessing options. Make sure that "Overwrite the outputs of geoprocessing operations" and "Add results of geoprocessing operations to the display" are selected.

- Use ArcCatalog to create the new toolbox, DiseaseIncidence.tbx, and add a model to it. Then, in ArcToolbox, right-click the ArcToolbox icon at the top of the ArcToolbox window, click Add Toolbox, and add your new toolbox.

- Your ModelBuilder model is a simple one. It has only two processes, each a kernel density estimation producing the two output rasters. *Hint 1:* Raster layer names cannot be longer than 13 characters. *Hint 2:* Right-click each output raster in the model and click Add To Display so that you don't have to add the outputs yourself.

- Symbolize the female mortality surface using Classified from the left panel of the Symbology tab of Layer Properties. Use the Standard Deviation Classification method with 1/3 standard deviation intervals. Then, for male mortality, import symbolization from the female mortality layer so that both layers have identical break points and color ramp. Then you can easily compare female and male incidence by turning one of them off and on while the other is on.

- *Hint:* If you need to rerun your model, click Model, Validate Entire Model.

Chapter 11 footnote references

1. Wayne D. Rosamond et al, "Trends in Incidence of Myocardial Infarction and in Mortality Due to Coronary Heart Disease, 1987 to 1994," *New England Journal of Medicine* 339, no. 13 (1998): 861–67.

2. William O'Hare and Mark Mather, "The Growing Number of Kids in Severely Distressed Neighborhoods: Evidence from the 2000 Census," In *Kids Count*, a publication of the Annie E. Casey Foundation and the Population Reference Bureau, revised October 2003, http://www.aecf.org/upload/publicationfiles/da3622h1280.pdf.

3. Robyn M. Dawes, "The Robust Beauty of Improper Linear Models in Decision Making," *American Psychologist* 7, no. 34 (1979): 571–82.

Appendix A

Task index

Software tool/concept, tutorial(s) in which it appears

3D Analyst extension, **10-1**

3D animation, **10-5**

3D Editor toolbar, **10-8**

3D effects, **10-6**

3D symbols, **10-7**

Add an aerial image, **10-3**

Add coverage, **5-5**

Add displacement links, **7-5**

Add elements to layout, **3-4**

Add a field, **3-7**

Add graph to layout, **3-6**

Add hyperlink, **3-1**

Add labels to model for documentation, **6-7**

Add layer, **1-2**

Add layer (.lyr) file, **2-3**

Add layer group, **2-8**

Add layers to group, **2-8**

Add model labels, **6-7**

Add model name and description for documentation, **6-7**

Add model parameters, **6-7**

Add raster maps from the Esri web service, **3-8**

Add raster maps from USGS, **5-10**

Add report to layout, **3-5**

Add variables to model, **6-7**

Add vertex points, **7-1**

Add x,y coordinates, **4-5**

Advanced sorting, **1-8**

Advanced time properties, **3-7**

Aggregate data, **2-7, 4-6**

Alias tables, **8-5**

Animation time properties, **3-7**

ArcCatalog utilities, **4-2**

ArcGIS 3D Analyst extension, **10-1**

ArcGIS Online, **3-8**

ArcGIS Spatial Analyst extension, **11-1**

ArcGlobe, **10-10**

ArcScene, **10-1**

Assign environmental settings, **11-3**

Assign projection, **5-2, 5-3, 5-4**

Attribute table, **1-8**

Auto Hide Catalog, **1-2**

Buffer lines, **9-1**

Buffer points, **9-1, 9-3**

Build file geodatabase, **4-1**

Build map animation, **3-7**

CAD drawing files, **5-5**

Calculate column, **4-3**

Calculate geometry, **4-5**

Centroid coordinates, **4-5**

Change map projection, **5-2, 5-3, 5-4**

Clear selected features, **1-7**

Clip features, **6-2**

Cluster analysis, **9-4**

Compress file geodatabase, **4-2**

Convert coverage to a feature class, **5-5**

Convert features to raster, **11-5**

Convert labels to annotation, **1-9**

Copy feature layers, **4-2**

Correct source addresses using interactive rematch, **8-1, 8-2**

Correct street reference layer addresses, **8-4**

Count geocoded records by ZIP Code, **8-1**

Create 3D scenes, **10-1**

Create 3D transparency effect, **10-6**

Create an address locator for streets with zone, **8-2**

Create an address locator for ZIP Codes, **8-2**

Create choropleth maps, **2-2, 2-3, 2-4, 2-5**

Create custom classes, **2-3**

Create feature class from XY table, **4-5**

Create file geodatabase, **4-1**

Create fly-through animations, **10-5**

Create hillshade for elevation, **11-2**

Create line feature, **7-2**

Create line of sight, **10-9**

Create model, **6-7, 11-6**

Create multiple 3D views, **10-4**

Create point feature, **7-3**

Create polygon feature, **7-1**

Create poverty contour, **11-6**

Create qualitative point and polygon maps, **2-1**

Create quantitative point and polygon maps, **2-2**

Create scatter plot, **9-3**

Create spatial bookmarks, **1-3**

Create toolbox, **11-6**

Create triangulated irregular network (TIN), **10-1**

Custom map layouts, **3-2**

Cut polygons tool, **7-4**

Data mining, **9-4**

Definition query, **2-1**

Delete an attribute column, **4-3**

Delete feature layers, **4-2**

Delete polygon, **7-1**

Delete vertex points, **7-1**

Density maps, **2-5**

Digitize line features, **7-2**

Digitize point features, **7-3**

Digitize polygon features, **7-1**

Dissolve features, **6-3**

Dot density maps, **2-6**

Download American Community Survey tables, **5-8**

Download Census TIGER/Line files, **5-6**

Download data from NationalAtlasnationalatlas. gov, **5-11**

Drag and drop a layer from Catalog, **1-2**

Drape features to TIN, **10-3**

Edit 3D attributes, **10-8**

Edit 3D objects, **10-8**

Edit displacement links, **7-5**

Edit feature attribute data, **7-1, 7-3**

Edit placement Placement tool, **10-8**

Edit tool, **7-4, 7-5**

Edit vertex points, **7-1**

Editor toolbar, **7-1**

Examine metadata, **4-2, 5-1**

Examine raster map layer, **11-1**

Explore ArcGlobe, **10-10**

Export 3D animation to video, **10-5**

Export feature class, **2-7**

Export image, **3-2**

Export layout, **3-2**

Export shapefiles to CAD, **5-5**

Export table, **4-5**

Extract by mask, **11-1**

Extract features using data queries, **6-1**

Extract raster value points, **11-4**

Field move, **1-8**

Field sort, **1-8**

Field statistics, **1-8**

File geodatabase, **4-1**

Find features, **1-5**

Fishnet maps, **2-7**

Fix and rematch ZIP Codes, **8-1**

Fixed zoom in, **1-3**

Fixed zoom out, **1-3**

Fly through a scene, **10-4**

Generalize tool, **7-4**

Geocode by streets, **8-2**

Geocode by ZIP Code, **8-1**

Geoprocessing options, **6-7, 11-6**

Graduated symbols, **2-5**

Gravity model, **9-3**

Hillshade, **11-2**

Hyperlinks, **3-1**

Identify problem streets using interactive rematch,
    8-3

Identify tool, **1-5**

Import a data table to a file geodatabase, **4-1**

Import a shapefile to a file geodatabase, **4-1**

Import layer files, **2-3**

Import text data into Microsoft Excel, **5-7, 5-8**

Interactive rematch addresses, **8-3**

Intersect layers, **6-5**

Join tables, **4-4**

Kernel density maps, **11-3, 11-6**

Label features, **1-9**

Label properties, **1-9**

Landform analysis, **10-9**

Layer color, **1-2, 2-4**

Layer display order, **1-2**

Layer groups, **2-8**

Layer package files (.lpk), **2-8**

Layout elements, **3-2**

Layout guidelines, **3-2**

Layout legends, **3-2**

Layout page orientation, **3-2**

Layout reports, **3-5**

Layout text, **3-2**

Layout title, **3-2**

Load 3D animation, **10-5**

Magnifier properties, **1-3**

Magnifier window, **1-3**

Manual classes, **2-3**

Manually change class colors, **2-4**

Map animation, **3-7**

Map document properties, **1-1**

Map projections, **5-2**

Map scale, **3-1**

Map tips, **3-1**

Measure button, **1-3**

Measure distances, **1-4**

Measurement units, **1-4**

Mercator projection, **5-2**

Merge features, **6-4**

Metadata, **5-1**

ModelBuilder, **6-7, 11-6**

Modify attribute table, **4-3**

Move a field, **1-8**

Move a polygon, **7-1**

Multiple data frames, **3-4**

Multiple output pages, **3-7**

Multiple ring buffer, **9-3**

Navigate a 3D scene, **10-4**

Normalized maps, **2-5**

Open a metadata file, **5-1**

Open and run a finished model, **6-7**

Overview window, **1-3**

Pan, **1-3**

Perform a line-of-sight analysis, **10-9**

Play an animation, **3-7, 10-5**

Preview layers, **4-2**

Process tabular data, **5-8, 5-7**

Query builder, **2-1, 6-6**

Query using reclassify, **11-5**

Raster analysis settings, **11-3, 11-4**

Raster queries, **11-5**

Rebuild a street locator, **8-4**

Reclassify tool, **11-5**

Record a 3D animation, **10-5**

Relative paths, **1-2**

Rematch interactively by correcting input addresses, **8-1, 8-3**

Rematch interactively by pointing on the map, **8-3**

Rematch interactively using an edited street segment, **8-4**

Remove a layer, **1-2**

Rename a layer, **2-1**

Reset a model, **6-7**

Rotate tool, **7-4**

Run a model, **6-7**

Save 3D animation, **10-5**

Save graph, **3-6**

Save layer file (.lyr), **2-3, 2-4**

Save map document, **1-1**

Save report, **3-5**

Scale bar, **3-2**

Scatterplots, **9-3**

Segment angles, **7-4**

Segment lengths, **7-4**

Select by attributes, **4-3, 6-1**

Select by graphic, **1-7**

Select by location, **6-2, 11-3**

Select elements, **1-9**

Select features, **1-6, 6-1**

Select records, **1-8**

Selectable layers, **1-6**

Selection color, **1-6**

Selection symbol, **1-6**

Set advanced time properties, **3-7**

Set geoprocessing options, **6-7, 11-4**

Set observer location, **10-4**

Set raster environment, **11-1**

Set snapping tools, **7-3**

Set target location, **10-4**

Set time properties, **3-7**

Set USA projections, **5-3**

Set world projections, **5-2**

Smooth tool, **7-4**

Snapping tools, **7-2**

Sort ascending, **1-8**

Sort descending, **1-8**

Sort field, **1-8**

Spatial joins, **4-6**

Spatially adjust features, **7-5**

State plane coordinate systems, **5-4**

Statistics, **1-8, 11-3**

Summarize a column, **4-5**

Switch selections, **1-8**

Symbolize a choropleth map, **2-1, 2-2**

Table queries, **9-1**

Time slider window, **3-7**

Trace tool, **7-4**

Transparency effect, **10-6**

Turn labels off, **1-9**

Turn labels on, **1-9**

Turn layer off, **1-2**

Turn layer on, **1-2**

Union features, **6-6**

Use advanced editing tools, **7-4**

Use alias table, **8-5**

Use data queries to extract features, **6-1**

Use web map service, **5-9, 5-10**

UTM coordinate system, **5-4**

Vector data formats, **5-5**

Visible scale range, **3-1**

Zoom in, **1-3, 3-1, 3-2, 10-2**

Zoom next extent, **1-3**

Zoom previous extent, **1-3**

Zoom selected features, **1-8, 6-1, 6-2**

Zoom to a bookmark, **1-3**

Zoom Whole Page, **3-2**

Appendix B

Data source credits

CMUCampus

\EsriPress\GIST1\Data\CMUCampus\CampusMap.dwg, courtesy of Carnegie Mellon University.

\EsriPress\GIST1\Data\CMUCampus\CampusMap.dwg\Annotation, courtesy of Carnegie Mellon University.

\EsriPress\GIST1\Data\CMUCampus\CampusMap.dwg\MultiPatch, courtesy of Carnegie Mellon University.

\EsriPress\GIST1\Data\CMUCampus\CampusMap.dwg\Point, courtesy of Carnegie Mellon University.

\EsriPress\GIST1\Data\CMUCampus\CampusMap.dwg\Polygon, courtesy of Carnegie Mellon University.

\EsriPress\GIST1\Data\CMUCampus\CampusMap.dwg\Polyline, courtesy of Carnegie Mellon University.

\EsriPress\GIST1\Data\CMUCampus\Walkways.dwg (from CampusMap.dwg), courtesy of Carnegie Mellon University.

\EsriPress\GIST1\Data\CMUCampus\Walkways.dwg\Annotation, courtesy of Carnegie Mellon University.

\EsriPress\GIST1\Data\CMUCampus\Walkways.dwg\MultiPatch, courtesy of Carnegie Mellon University.

\EsriPress\GIST1\Data\CMUCampus\Walkways.dwg\Point, courtesy of Carnegie Mellon University.

\EsriPress\GIST1\Data\CMUCampus\Walkways.dwg\Polygon, courtesy of Carnegie Mellon University.

\EsriPress\GIST1\Data\CMUCampus\Walkways.dwg\Polyline, courtesy of Carnegie Mellon University.

\EsriPress\GIST1\Data\CMUCampus\CMUCampus.tif, image courtesy of US Geological Survey, Department of the Interior/USGS.

\EsriPress\GIST1\Data\CMUCampus\CMUCampus.tif\Band_1, image courtesy of US Geological Survey, Department of the Interior/USGS.

\EsriPress\GIST1\Data\CMUCampus\CMUCampus.tif\Band_2, image courtesy of US Geological Survey, Department of the Interior/USGS.

\EsriPress\GIST1\Data\CMUCampus\CMUCampus.tif\Band_3, image courtesy of US Geological Survey, Department of the Interior/USGS.

\EsriPress\GIST1\Data\CMUCampus\HBH.shp, courtesy of Carnegie Mellon University.

\EsriPress\GIST1\Data\CMUCampus\MainCampusBuildings.shp (from CampusMap.dwg), courtesy of Carnegie Mellon University.

DataFiles

\EsriPress\GIST1\Data\DataFiles\AllCoEdAttain.xlsx, courtesy of US Census Bureau.

\EsriPress\GIST1\Data\DataFiles\AllCoEdAttain.xlsc\HighEdAttainment$, courtesy of US Census Bureau.

\EsriPress\GIST1\Data\DataFiles\AlleghenyCountyBlockGroups.shp, courtesy of US Census Bureau.

\EsriPress\GIST1\Data\DataFiles\AlleghenyCountyTracts.shp, courtesy of US Census Bureau.

\EsriPress\GIST1\Data\DataFiles\AsianTop10States.xlsx, courtesy of US Census Bureau and Wilpen Gorr.

\EsriPress\GIST1\Data\DataFiles\AsianTop10States.xlsx\AsianTop10States$, courtesy of US Census Bureau and Wilpen Gorr.

\EsriPress\GIST1\Data\DataFiles\AutoTheftCrimeSeries.shp, courtesy of Department of City Planning, City of Pittsburgh.

\EsriPress\GIST1\Data\DataFiles\DCTract.shp, courtesy of US Census Bureau.

\EsriPress\GIST1\Data\DataFiles\Earthquakes.dbf, from Esri Data & Maps, 2004, courtesy of National Atlas of the United States, USGS.

\EsriPress\GIST1\Data\DataFiles\FishnetCalculations.xlsx, by Wilpen Gorr.

\EsriPress\GIST1\Data\DataFiles\PghCrosswalks.dbf, courtesy of Department of City Planning, City of Pittsburgh, and Wilpen Gorr.

\EsriPress\GIST1\Data\DataFiles\PghTracts.shp, courtesy of US Census Bureau.

\EsriPress\GIST1\Data\DataFiles\PittsburghSchools.csv, from Esri Data & Maps 2010, courtesy of US Census Bureau.

\EsriPress\GIST1\Data\DataFiles\Statistics.txt, courtesy of US Census Bureau and Wilpen Gorr.

\EsriPress\GIST1\Data\DataFiles\tl_2010_42003_tract10.shp, courtesy of US Census Bureau.

\EsriPress\GIST1\Data\DataFiles\Tornadoes2000.shp, courtesy of National Atlas of the United States, downloaded from NationalAtlas.gov (http://www.nationalatlas.gov/atlasftp.html#tornadx).

\EsriPress\GIST1\Data\DataFiles\Neighborhoods.zip, courtesy of Department of City Planning, City of Pittsburgh.

MaricopaCounty

\EsriPress\GIST1\Data\MaricopaCounty\CensusData.xlsx, courtesy of US Census Bureau.

\EsriPress\GIST1\Data\MaricopaCounty\CensusData.xlsx\CensusData,courtesy of US Census Bureau.

\EsriPress\GIST1\Data\MaricopaCounty\CensusData.xlsx\CensusData$, courtesy of US Census Bureau.

\EsriPress\GIST1\Data\MaricopaCounty\tl_2010_04013_cousub10.shp, courtesy of US Census Bureau.

\EsriPress\GIST1\Data\MaricopaCounty\tl_2010_04013_tract10.shp, courtesy of US Census Bureau.

EastLiberty

\EsriPress\GIST1\Data\Pittsburgh\EastLiberty\Building\Annotation, courtesy of Department of City Planning, City of Pittsburgh.

\EsriPress\GIST1\Data\Pittsburgh\EastLiberty\Building\Arc, courtesy of Department of City Planning, City of Pittsburgh.

\EsriPress\GIST1\Data\Pittsburgh\EastLiberty\Building\Label, courtesy of Department of City Planning, City of Pittsburgh.

\EsriPress\GIST1\Data\Pittsburgh\EastLiberty\Building\Polygon, courtesy of Department of City Planning, City of Pittsburgh.

\EsriPress\GIST1\Data\Pittsburgh\EastLiberty\Building\Tic, courtesy of Department of City Planning, City of Pittsburgh.

\EsriPress\GIST1\Data\Pittsburgh\EastLiberty\Curbs\Annotation, courtesy of Department of City Planning, City of Pittsburgh.

\EsriPress\GIST1\Data\Pittsburgh\EastLiberty\Curbs\Arc, courtesy of Department of City Planning, City of Pittsburgh.

\EsriPress\GIST1\Data\Pittsburgh\EastLiberty\Curbs\Tic, courtesy of Department of City Planning, City of Pittsburgh.

\EsriPress\GIST1\Data\Pittsburgh\EastLiberty\EastLib\Annotation, courtesy of Department of City Planning, City of Pittsburgh.

\EsriPress\GIST1\Data\Pittsburgh\EastLiberty\EastLib\Arc, courtesy of Department of City Planning, City of Pittsburgh.

\EsriPress\GIST1\Data\Pittsburgh\EastLiberty\EastLib\Label, courtesy of Department of City Planning, City of Pittsburgh.

\EsriPress\GIST1\Data\Pittsburgh\EastLiberty\EastLib\Polygon, courtesy of Department of City Planning, City of Pittsburgh.

\EsriPress\GIST1\Data\Pittsburgh\EastLiberty\EastLib\Tic, courtesy of Department of City Planning, City of Pittsburgh.

\EsriPress\GIST1\Data\Pittsburgh\EastLiberty\Parcel\Annotation, courtesy of Department of City Planning, City of Pittsburgh.

\EsriPress\GIST1\Data\Pittsburgh\EastLiberty\Parcel\Arc, courtesy of Department of City Planning, City of Pittsburgh.

\EsriPress\GIST1\Data\Pittsburgh\EastLiberty\Parcel\Label, courtesy of Department of City Planning, City of Pittsburgh.

\EsriPress\GIST1\Data\Pittsburgh\EastLiberty\Parcel\Polygon, courtesy of Department of City Planning, City of Pittsburgh.

\EsriPress\GIST1\Data\Pittsburgh\EastLiberty\Parcel\Tic, courtesy of Department of City Planning, City of Pittsburgh.

Shapefiles

\EsriPress\GIST1\Data\Pittsburgh\Shapefiles\Larcenies.shp, courtesy of Pittsburgh Bureau of Police.

\EsriPress\GIST1\Data\Pittsburgh\Shapefiles\Parks.shp, courtesy of Southwestern Pennsylvania Commission.

\EsriPress\GIST1\Data\Pittsburgh\Shapefiles\PittsburghSeriousCrimes2008.shp, Courtesy of Pittsburgh Bureau of Police.

\EsriPress\GIST1\Data\Pittsburgh\Shapefiles\PovertyTracts.xlsx, courtesy of US Census Bureau.

\EsriPress\GIST1\Data\Pittsburgh\Shapefiles\PovertyTracts.xlsx\PghPovertyTracts$ courtesy of US Census Bureau.

CBD

\EsriPress\GIST1\Data\Pittsburgh\CBD.gdb\BldgNameAlias, courtesy of Wilpen Gorr, Carnegie Mellon University.

\EsriPress\GIST1\Data\Pittsburgh\CBD.gdb\Bldgs, courtesy of Department of City Planning, City of Pittsburgh.

\EsriPress\GIST1\Data\Pittsburgh\CBD.gdb\Clients, courtesy of Kristen Kurland, Carnegie Mellon University.

\EsriPress\GIST1\Data\Pittsburgh\CBD.gdb\Curbs, courtesy of Department of City Planning, City of Pittsburgh.

\EsriPress\GIST1\Data\Pittsburgh\CBD.gdb\Histpnts, courtesy of Department of City Planning, City of Pittsburgh.

\EsriPress\GIST1\Data\Pittsburgh\CBD.gdb\Histsites, courtesy of Department of City Planning, City of Pittsburgh.

\EsriPress\GIST1\Data\Pittsburgh\CBD.gdb\Outline, courtesy of Department of City Planning, City of Pittsburgh.

\EsriPress\GIST1\Data\Pittsburgh\CBD.gdb\Parcels, courtesy of Department of City Planning, City of Pittsburgh.

\EsriPress\GIST1\Data\Pittsburgh\CBD.gdb\Streets, courtesy of Department of City Planning, City of Pittsburgh.

City

\EsriPress\GIST1\Data\Pittsburgh\City.gdb\ATMRobberies, courtesy of Pittsburgh Bureau of Police.

\EsriPress\GIST1\Data\Pittsburgh\City.gdb\ImmigrantBusinesses, courtesy of Kristen Kurland, Carnegie Mellon University.

\EsriPress\GIST1\Data\Pittsburgh\City.gdb\Neighborhoods, courtesy of Department of City Planning, City of Pittsburgh.

\EsriPress\GIST1\Data\Pittsburgh\City.gdb\PghStreets, courtesy of Department of City Planning, City of Pittsburgh.

\EsriPress\GIST1\Data\Pittsburgh\City.gdb\PghTracts, courtesy of US Census Bureau.

\EsriPress\GIST1\Data\Pittsburgh\City.gdb\Pittsburgh, courtesy of Southwestern Pennsylvania Commission.

\EsriPress\GIST1\Data\Pittsburgh\City.gdb\Rivers, courtesy of US Census Bureau.

\EsriPress\GIST1\Data\Pittsburgh\City.gdb\Schools, courtesy of Department of City Planning, City of Pittsburgh.

\EsriPress\GIST1\Data\Pittsburgh\City.gdb\Zoning, courtesy of Department of City Planning, City of Pittsburgh.

Midhill

\EsriPress\GIST1\Data\Pittsburgh\Midhill.gdb\Bldgs, courtesy of Department of City Planning, City of Pittsburgh.

\EsriPress\GIST1\Data\Pittsburgh\Midhill.gdb\CADCalls, courtesy of Department of City Planning, City of Pittsburgh.

\EsriPress\GIST1\Data\Pittsburgh\Midhill.gdb\CommercialProperties, courtesy of Department of City Planning, City of Pittsburgh.

\EsriPress\GIST1\Data\Pittsburgh\Midhill.gdb\Curbs, courtesy of Department of City Planning, City of Pittsburgh.

\EsriPress\GIST1\Data\Pittsburgh\Midhill.gdb\Outline, courtesy of Department of City Planning, City of Pittsburgh.

\EsriPress\GIST1\Data\Pittsburgh\Midhill.gdb\Schools, courtesy of National Agriculture Imagery Program (NAIP), US Department of Agriculture, Farm Service Agency.

\EsriPress\GIST1\Data\Pittsburgh\Midhill.gdb\Streets, courtesy of Department of City Planning, City of Pittsburgh.

Phipps

\EsriPress\GIST1\Data\Pittsburgh\Phipps.gdb\Bldgs, courtesy of Department of City Planning, City of Pittsburgh.

\EsriPress\GIST1\Data\Pittsburgh\Phipps.gdb\Curbs, courtesy of Department of City Planning, City of Pittsburgh.

\EsriPress\GIST1\Data\Pittsburgh\Phipps.gdb\Image\Band_1, courtesy of US Geological Survey, Department of the Interior/USGS.

\EsriPress\GIST1\Data\Pittsburgh\Phipps.gdb\Image\Band_2, courtesy of US Geological Survey, Department of the Interior/USGS.

\EsriPress\GIST1\Data\Pittsburgh\Phipps.gdb\Image\Band_3, courtesy of US Geological Survey, Department of the Interior/USGS.

\EsriPress\GIST1\Data\Pittsburgh\Phipps.gdb\PhippsAddition, digitized by Kristen Kurland, Carnegie Mellon University.

\EsriPress\GIST1\Data\Pittsburgh\Phipps.gdb\Topo, courtesy of Department of City Planning, City of Pittsburgh.

\EsriPress\GIST1\Data\Pittsburgh\PublicPools.gdb\JackStackVisits, courtesy of Pittsburgh Citiparks Department.

\EsriPress\GIST1\Data\Pittsburgh\PublicPools.gdb\PittsburghBlockCentroids, Courtesy of US Census Bureau and Wilpen Gorr

\EsriPress\GIST1\Data\Pittsburgh\PublicPools.gdb\Pools, courtesy of Pittsburgh Citiparks Department.

\EsriPress\GIST1\Data\Pittsburgh\PublicPools.gdb\Pooltags, courtesy of Pittsburgh Citiparks Department.

Zone2

\EsriPress\GIST1\Data\Pittsburgh\Zone2.gdb\Outline, courtesy of Department of City Planning, City of Pittsburgh.

\EsriPress\GIST1\Data\Pittsburgh\Zone2.gdb\Streets, courtesy of US Census Bureau.

RochesterNY

\EsriPress\GIST1\Data\RochesterNY\LakePrecinct.gdb\lakebars, courtesy of Rochester, New York, Police Department.

\EsriPress\GIST1\Data\RochesterNY\LakePrecinct.gdb\LakeBusinesses, courtesy of InfoUSA.

\EsriPress\GIST1\Data\RochesterNY\LakePrecinct.gdb\lakecarbeats, courtesy of Rochester, New York, Police Department.

\EsriPress\GIST1\Data\RochesterNY\LakePrecinct.gdb\lakeprecinct, courtesy of Rochester, New York, Police Department.

\EsriPress\GIST1\Data\RochesterNY\LakePrecinct.gdb\lakestreets, courtesy of Rochester, New York, Police Department.

\EsriPress\GIST1\Data\RochesterNY\RochesterPolice.gdb\business, courtesy of Rochester, New York, Police Department.

\EsriPress\GIST1\Data\RochesterNY\RochesterPolice.gdb\carbeats, courtesy of Rochester, New York, Police Department.

\EsriPress\GIST1\Data\RochesterNY\RochesterPolice.gdb\SIC, courtesy US Census Bureau.

\EsriPress\GIST1\Data\RochesterNY\RochesterPolice.gdb\streets, courtesy US Census Bureau.

SpatialAnalyst

\EsriPress\GIST1\Data\SpatialAnalyst\LandUse\LandUse_Pgh.tif, image courtesy of US Geological Survey, Department of the Interior/USGS.

\EsriPress\GIST1\Data\SpatialAnalyst\LandUse\LandUse_Pgh.prj, courtesy of Wilpen Gorr.

\EsriPress\GIST1\Data\SpatialAnalyst\SpatialAnalyst.gdb\AllCoBlkGrps, courtesy of US Census Bureau TIGER.

\EsriPress\GIST1\Data\SpatialAnalyst\SpatialAnalyst.gdb\AllCoBlocks, courtesy of US Census Bureau TIGER.

\EsriPress\GIST1\Data\SpatialAnalyst\SpatialAnalyst.gdb\NED, courtesy of US Geological Survey, Department of the Interior/USGS.

\EsriPress\GIST1\Data\SpatialAnalyst\SpatialAnalyst.gdb\OHCA, courtesy of Children's Hospital of Pittsburgh.

\EsriPress\GIST1\Data\SpatialAnalyst\SpatialAnalyst.gdb\PennHills, courtesy of US Census Bureau.

\EsriPress\GIST1\Data\SpatialAnalyst\SpatialAnalyst.gdb\Pittsburgh, courtesy of US Census Bureau.

\EsriPress\GIST1\Data\SpatialAnalyst\SpatialAnalyst.gdb\Rivers, courtesy of US Census Bureau.

\EsriPress\GIST1\Data\SpatialAnalyst\SpatialAnalyst.gdb\Shaler, courtesy of US Census Bureau.

\EsriPress\GIST1\Data\SpatialAnalyst\SpatialAnalyst.gdb\ZoningCommercialBuffer, courtesy of Department of City Planning, City of Pittsburgh.

\EsriPress\GIST1\Data\SpatialAnalyst\LandUse.lyr, layer file courtesy of Wilpen Gorr with symbolization for LandUse_Pgh.tif.

3DAnalyst

\EsriPress\GIST1\Data\3DAnalyst.gdb\Bldgs, courtesy of Department of City Planning, City of Pittsburgh.

\EsriPress\GIST1\Data\3DAnalyst.gdb\Curbs, courtesy of Department of City Planning, City of Pittsburgh.

\EsriPress\GIST1\Data\3DAnalyst.gdb\Image\Band_1, courtesy of US Geological Survey, Department of the Interior/USGS.

\EsriPress\GIST1\Data\3DAnalyst.gdb\Image\Band_2, courtesy of US Geological Survey, Department of the Interior/USGS.

\EsriPress\GIST1\Data\3DAnalyst.gdb\NorthShoreTrees, courtesy of Kristen Kurland, Carnegie Mellon University.

\EsriPress\GIST1\Data\3DAnalyst.gdb\NorthShoreVehicles, courtesy of Kristen Kurland, Carnegie Mellon University.

\EsriPress\GIST1\Data\3DAnalyst.gdb\ProposedBldgs, courtesy of Kristen Kurland, Carnegie Mellon University.

\EsriPress\GIST1\Data\3DAnalyst.gdb\Rivers, courtesy of US Census Bureau.

\EsriPress\GIST1\Data\3DAnalyst.gdb\StudyArea, courtesy of Kristen Kurland, Carnegie Mellon University.

\EsriPress\GIST1\Data\3DAnalyst.gdb\Topo, courtesy of Department of City Planning, City of Pittsburgh.

AlleghenyCounty

\EsriPress\GIST1\Data\AlleghenyCounty.gdb\CountySchools, from Esri Data & Maps 2010, courtesy of US Census Bureau.

\EsriPress\GIST1\Data\AlleghenyCounty.gdb\Munic, courtesy of Southwestern Pennsylvania Commission.

\EsriPress\GIST1\Data\AlleghenyCounty.gdb\Parks, courtesy of Southwestern Pennsylvania Commission.

\EsriPress\GIST1\Data\AlleghenyCounty.gdb\Rivers, courtesy of US Census Bureau.

\EsriPress\GIST1\Data\AlleghenyCounty.gdb\Tracts2010, courtesy of US Census Bureau.

NYC

\EsriPress\GIST1\Data\NYC.gdb\Boroughs. Material titled GIS Data used with permission of New York City Department of City Planning. All rights reserved.

\EsriPress\GIST1\Data\NYC.gdb\BronxWater, courtesy of US Census Bureau.

\EsriPress\GIST1\Data\NYC.gdb\BronxWaterfrontParks. Material titled GIS Data used with permission of New York City Department of City Planning. All rights reserved.

\EsriPress\GIST1\Data\NYC.gdb\BrooklynWaterfrontParks. Material titled GIS Data used with permission of New York City Department of City Planning. All rights reserved.

\EsriPress\GIST1\Data\NYC.gdb\CouncilDistricts. Material titled GIS Data used with permission of New York City Department of City Planning. All rights reserved.

\EsriPress\GIST1\Data\NYC.gdb\Facilities. Material titled GIS Data used with permission of New York City Department of City Planning. All rights reserved.

\EsriPress\GIST1\Data\NYC.gdb\FireCompanies. Material titled GIS Data used with permission of New York City Department of City Planning. All rights reserved.

\EsriPress\GIST1\Data\NYC.gdb\HealthAreas. Material titled GIS Data used with permission of New York City Department of City Planning. All rights reserved.

\EsriPress\GIST1\Data\NYC.gdb\HealthCenters. Material titled GIS Data used with permission of New York City Department of City Planning. All rights reserved.

\EsriPress\GIST1\Data\NYC.gdb\KingsWater, courtesy of US Census Bureau.

\EsriPress\GIST1\Data\NYC.gdb\ManhattanStreets. Material titled GIS Data used with permission of New York City Department of City Planning. All rights reserved.

\EsriPress\GIST1\Data\NYC.gdb\ManhattanTracts, courtesy of US Census Bureau.

\EsriPress\GIST1\Data\NYC.gdb\ManhattanWaterfrontParks. Material titled GIS Data used with permission of New York City Department of City Planning. All rights reserved.

\EsriPress\GIST1\Data\NYC.gdb\MetroRoads, from Esri Data & Maps, 2004, courtesy of US Census Bureau.

\EsriPress\GIST1\Data\NYC.gdb\Neighborhoods. Material titled GIS Data used with permission of New York City Department of City Planning. All rights reserved.

\EsriPress\GIST1\Data\NYC.gdb\NewYorkWater, courtesy of US Census Bureau.

\EsriPress\GIST1\Data\NYC.gdb\PolicePrecincts. Material titled GIS Data used with permission of New York City Department of City Planning. All rights reserved.

\EsriPress\GIST1\Data\NYC.gdb\QueensWater, courtesy of US Census Bureau.

\EsriPress\GIST1\Data\NYC.gdb\QueensWaterfrontParks. Material titled GIS Data used with permission of New York City Department of City Planning. All rights reserved.

\EsriPress\GIST1\Data\NYC.gdb\RichmondWater, courtesy of US Census Bureau.

\EsriPress\GIST1\Data\NYC.gdb\SchoolDistricts. Material titled GIS Data used with permission of New York City Department of City Planning. All rights reserved.

\EsriPress\GIST1\Data\NYC.gdb\StatenIslandWaterfrontParks. Material titled GIS Data used with permission of New York City Department of City Planning. All rights reserved.

\EsriPress\GIST1\Data\NYC.gdb\Water, courtesy of US Census Bureau.

\EsriPress\GIST1\Data\NYC.gdb\ZoningLandUse. Material titled GIS Data used with permission of New York City Department of City Planning. All rights reserved.

UnitedStates

\EsriPress\GIST1\Data\UnitedStates.gdb\CACounties, from Esri Data & Maps 2010, courtesy of ArcUSA, US Census Bureau, Esri (POP2010 field).

\EsriPress\GIST1\Data\UnitedStates.gdb\CAOrangeCountyTracts, from Esri Data & Maps 2010, courtesy of Tele Atlas, US Census Bureau, Esri (POP2010 field).

\EsriPress\GIST1\Data\UnitedStates.gdb\COCounties, Esri Data & Maps 2010, courtesy of ArcUSA, US Census Bureau, Esri (POP2010 field).

\EsriPress\GIST1\Data\UnitedStates.gdb\Congress111, courtesy of US Census Bureau.

\EsriPress\GIST1\Data\UnitedStates.gdb\COStreets, Esri Data & Maps 2010, courtesy of ArcUSA, US Census Bureau, Esri (POP2010 field).

\EsriPress\GIST1\Data\UnitedStates.gdb\COStreets1, from Esri Data & Maps 2010, courtesy of ArcUSA, US Census Bureau, Esri (POP2010 field).

\EsriPress\GIST1\Data\UnitedStates.gdb\COStreets2, from Esri Data & Maps 2010, courtesy of ArcUSA, US Census Bureau, Esri (POP2010 field).

\EsriPress\GIST1\Data\UnitedStates.gdb\COUrban1, courtesy of US Census Bureau.

\EsriPress\GIST1\Data\UnitedStates.gdb\COUrban2, courtesy of US Census Bureau.

\EsriPress\GIST1\Data\UnitedStates.gdb\Water1, courtesy of US Census Bureau.

\EsriPress\GIST1\Data\United States.gdb\Water2, courtesy of US Census Bureau.

\EsriPress\GIST1\Data\UnitedStates.gdb\HHWZIPCodes, courtesy of Pennsylvania Resources Council.

\EsriPress\GIST1\Data\UnitedStates.gdb\MilitaryBnd, courtesy of DISDI, downloaded from Data.gov (http://www.data.gov/geodata/g735939/), DISDI. Data.gov and the US federal government cannot vouch for the data or analyses derived from this data after the data has been retrieved from Data.gov.

\EsriPress\GIST1\Data\UnitedStates.gdb\MilitaryPt, courtesy of DISDI, downloaded from Data.gov (http://www.data.gov/geodata/g736859/), DISDI. Data.gov and the Federal Government cannot vouch for the data or analyses derived from this data after the data has been retrieved from Data.gov.

\EsriPress\GIST1\Data\UnitedStates.gdb\PACities, from Esri Data & Maps 2010, courtesy of National Atlas of the United States.

\EsriPress\GIST1\Data\UnitedStates.gdb\PACounties, from Esri Data & Maps 2010, courtesy of ArcUSA, US Census Bureau, Esri (POP2010 field).

\EsriPress\GIST1\Data\UnitedStates.gdb\PAPublicSchools, from Esri Data & Maps 2010, courtesy of Tele Atlas (TomTom), US Census Bureau, Esri (POP2010 field).

\EsriPress\GIST1\Data\UnitedStates.gdb\PATracts, from Esri Data & Maps 2010,courtesy of Tele Atlas (TomTom), US Census Bureau, Esri (POP2010 field).

\EsriPress\GIST1\Data\UnitedStates.gdb\PAZIP, courtesy of Tele Atlas (TomTom), Esri (POP2010 field).

\EsriPress\GIST1\Data\UnitedStates.gdb\USCities, from Esri Data & Maps 2010, courtesy of US Census Bureau.

\EsriPress\GIST1\Data\UnitedStates.gdb\USCities_dtl, from Esri Data & Maps 2010, courtesy of US Census Bureau.

\EsriPress\GIST1\Data\UnitedStates.gdb\USCounties, from Esri Data & Maps 2010, courtesy of ArcUSA, US Census Bureau, Esri (POP2010 field).

\EsriPress\GIST1\Data\UnitedStates.gdb\USStates, from Esri Data & Maps 2010, courtesy of ArcUSA, US Census Bureau, Esri (POP2010 field).

World

\EsriPress\GIST1\Data\World.gdb\Country, courtesy of *ArcWorld Supplement*.

\EsriPress\GIST1\Data\World.gdb\Ocean, from Esri Data & Maps, 2004, courtesy of Esri.

MyExercises\FinishedExercises

\EsriPress\GIST1\MyExercises\FinishedExercises\Chapter5\Chapter5.gdb\tl_2010_17031_cousub10.zip, courtesy of US Census Bureau.

\EsriPress\GIST1\MyExercises\FinishedExercises\Chapter5\Topo.DWG, courtesy of Department of City Planning, City of Pittsburgh.

\EsriPress\GIST1\MyExercises\FinishedExercises\Chapter5\Chapter5.gdb\ACSLessThanHighSch, courtesy of US Census Bureau.

\EsriPress\GIST1\MyExercises\FinishedExercises\Chapter5\Chapter5.gdb\AfricanizedBees, courtesy of Data.gov.

\EsriPress\GIST1\MyExercises\FinishedExercises\Chapter5\Chapter5.gdb\Building, courtesy of Department of City Planning, City of Pittsburgh.

\EsriPress\GIST1\MyExercises\FinishedExercises\Chapter5\Chapter5.gdb\EcoRegions, courtesy of Data.gov.

\EsriPress\GIST1\MyExercises\FinishedExercises\Chapter5\Chapter5.gdb\Tornadoes, courtesy of Data.gov.

\EsriPress\GIST1\MyExercises\FinishedExercises\Chapter5\Chapter5.gdb\TractsPop2010, courtesy of US Census Bureau.

\EsriPress\GIST1\MyExercises\FinishedExercises\Chapter5\Chapter5.gdb\Chaptertl_2010_17031_areawater.zip, courtesy of US Census Bureau.

\EsriPress\GIST1\MyExercises\FinishedExercises\Chapter5\Chapter5.gdb\tl_2010_17031_cousub10.zip, courtesy of US Census Bureau.

\EsriPress\GIST1\MyExercises\FinishedExercises\Chapter5\Chapter5.gdb\tl_2010_17031_tract10.zip, courtesy of US Census Bureau.

\EsriPress\GIST1\MyExercises\FinishedExercises\Chapter5\DataGovShapefiles\afrbeep020.shp, courtesy of National Atlas of the United States, downloaded from NationalAtlas.gov (http://nationalatlas. gov/atlasftp.html#afrbeep).

\EsriPress\GIST1\MyExercises\FinishedExercises\Chapter5\DataGovShapefiles\ecoregp075.shp, courtesy of National Atlas of the United States, downloaded from NationalAtlas.gov (http://nationalatlas. gov/atlasftp.html#ecoregp).

\EsriPress\GIST1\MyExercises\FinishedExercises\Chapter5\DataGovShapefiles\tornadx020.shp, courtesy of National Atlas of the United States, downloaded from NationalAtlas.gov (http://www. nationalatlas.gov/atlasftp.html#tornadx).

\EsriPress\GIST1\MyExercises\FinishedExercises\Chapter5\ACS_10_5YR_S1501.zip, courtesy of US Census Bureau.

\EsriPress\GIST1\MyExercises\FinishedExercises\Chapter5\DEC_10_SF2_QTP1.zip, courtesy of US Census Bureau.

Supplemental figure credits

Chapter 3

Screen capture from ArcGIS Online, courtesy of i-cubed, information integration + imaging, LLC— distributed through i-cubed's DataDoors Archive Management www.datadoors.net.

Screen captures from ArcGIS Online, courtesy of National Agriculture Imagery Program, US Department of Agriculture, Farm Service Agency.

Screen captures from ArcGIS Online, courtesy of TomTom MultiNet.

Screen captures of www.census.gov, courtesy of US Census Bureau.

Screen captures of http://seamless.usgs.gov, courtesy of USGS.

Chapter 5

Screen captures of http://nationalatlas.gov/mld/afrbeep.html, courtesy of Data.gov. Data.gov and the Federal Government cannot vouch for the data or analyses derived from this data after the data has been retrieved from Data.gov.

Screen captures from ArcGIS Online courtesy of i-cubed, information integration + imaging, LLC—distributed through i-cubed's DataDoors Archive Management www.datadoors.net.

Screen captures of ArcGlobe, data sources:

Elevation (30m)—Source: USGS. The data is from the National Elevation Dataset (NED) produced by the United States Geological Survey (USGS).

Elevation (90m/1km)—Source: NASA, National Geospatial-Intelligence Agency (NGA), USGS. The data is from the National Elevation Dataset (NED) produced by the United States Geological Survey (USGS).

Copyright:© 2009 Esri, i-cubed, GeoEye elevation data includes 90m Shuttle Radar Topography Mission (SRTM) elevation data from NASA and NGA where it is available and 1km GTOPO30 data from the USGS elsewhere.

Imagery—Copyright:© 2009 Esri, i-cubed, GeoEye. This globe presents low-resolution imagery for the world and high-resolution imagery for the United States and other metropolitan areas around the world. The globe includes NASA Blue Marble: Next Generation 500m resolution imagery at small scales (above 1:1,000,000), i-cubed 15m eSAT imagery at medium-to-large scales (down to 1:70,000) for the world, and USGS 15m Landsat imagery for Antarctica. It also includes 1m i-cubed Nationwide Select imagery for the continental United States, and GeoEye IKONOS 1m resolution imagery for Hawaii, parts of Alaska, and several hundred metropolitan areas around the world.

Boundaries and Places—Copyright:© 2009 Esri, AND, TANA. The map was developed by Esri using administrative and cities data from Esri and AND Mapping for the world and Tele Atlas administrative, cities, and landmark data for North America and Europe.

Transportation—Copyright:© 2009 Esri, AND, TANA. The map was developed by Esri using Esri highway data, National Geospatial-Intelligence Agency (NGA) airport data, AND road and railroad data for the world, and Tele Atlas.

Appendix C

Data license agreement

Important:
Read carefully before opening the sealed media package

Environmental Systems Research Institute Inc. (Esri) is willing to license the enclosed data and related materials to you only upon the condition that you accept all of the terms and conditions contained in this license agreement. Please read the terms and conditions carefully before opening the sealed media package. By opening the sealed media package, you are indicating your acceptance of the Esri License Agreement. If you do not agree to the terms and conditions as stated, then Esri is unwilling to license the data and related materials to you. In such event, you should return the media package with the seal unbroken and all other components to Esri.

Esri license agreement

This is a license agreement, and not an agreement for sale, between you (Licensee) and Environmental Systems Research Institute, Inc. (Esri). This Esri License Agreement (Agreement) gives Licensee certain limited rights to use the data and related materials (Data and Related Materials). All rights not specifically granted in this Agreement are reserved to Esri and its Licensors.

Reservation of Ownership and Grant of License: Esri and its Licensors retain exclusive rights, title, and ownership to the copy of the Data and Related Materials licensed under this Agreement and, hereby, grant to Licensee a personal, nonexclusive, nontransferable, royalty-free, worldwide license to use the Data and Related Materials based on the terms and conditions of this Agreement. Licensee agrees to use reasonable effort to protect the Data and Related Materials from unauthorized use, reproduction, distribution, or publication.

Proprietary Rights and Copyright: Licensee acknowledges that the Data and Related Materials are proprietary and confidential property of Esri and its Licensors and are protected by United States copyright laws and applicable international copyright treaties and/or conventions.

Permitted Uses: Licensee may install the Data and Related Materials onto permanent storage device(s) for Licensee's own internal use.

Licensee may make only one (1) copy of the original Data and Related Materials for archival purposes during the term of this Agreement unless the right to make additional copies is granted to Licensee in writing by Esri.

Licensee may internally use the Data and Related Materials provided by Esri for the stated purpose of GIS training and education.

Uses Not Permitted: Licensee shall not sell, rent, lease, sublicense, lend, assign, time-share, or transfer, in whole or in part, or provide unlicensed Third Parties access to the Data and Related Materials or portions of the Data and Related Materials, any updates, or Licensee's rights under this Agreement.

Licensee shall not remove or obscure any copyright or trademark notices of Esri or its Licensors.

Term and Termination: The license granted to Licensee by this Agreement shall commence upon the acceptance of this Agreement and shall continue until such time that Licensee elects in writing to discontinue use of the Data or Related Materials and terminates this Agreement. The Agreement shall automatically terminate without notice if Licensee fails to comply with any provision of this Agreement. Licensee shall then return to Esri the Data and Related Materials. The parties hereby agree that all provisions that operate to protect the rights of Esri and its Licensors shall remain in force should breach occur.

Disclaimer of Warranty: The Data and Related Materials contained herein are provided "as-is," without warranty of any kind, either express or implied, including, but not limited to, the implied warranties of merchantability, fitness for a particular purpose, or noninfringement. Esri does not warrant that the Data and Related Materials will meet Licensee's needs or expectations, that the use of the Data and Related Materials will be uninterrupted, or that all nonconformities, defects, or errors can or will be corrected. Esri is not inviting reliance on the Data or Related Materials for commercial planning or analysis purposes, and Licensee should always check actual data.

Data Disclaimer: The Data used herein has been derived from actual spatial or tabular information. In some cases, Esri has manipulated and applied certain assumptions, analyses, and opinions to the Data solely for educational training purposes. Assumptions, analyses, opinions applied, and actual outcomes may vary. Again, Esri is not inviting reliance on this Data, and the Licensee should always verify actual Data and exercise their own professional judgment when interpreting any outcomes.

Limitation of Liability: Esri shall not be liable for direct, indirect, special, incidental, or consequential damages related to Licensee's use of the Data and Related Materials, even if Esri is advised of the possibility of such damage.

No Implied Waivers: No failure or delay by Esri or its Licensors in enforcing any right or remedy under this Agreement shall be construed as a waiver of any future or other exercise of such right or remedy by Esri or its Licensors.

Order for Precedence: Any conflict between the terms of this Agreement and any FAR, DFAR, purchase order, or other terms shall be resolved in favor of the terms expressed in this Agreement, subject to the government's minimum rights unless agreed otherwise.

Export Regulation: Licensee acknowledges that this Agreement and the performance thereof are subject to compliance with any and all applicable United States laws, regulations, or orders relating to the export of data thereto. Licensee agrees to comply with all laws, regulations, and orders of the United States in regard to any export of such technical data.

Severability: If any provision(s) of this Agreement shall be held to be invalid, illegal, or unenforceable by a court or other tribunal of competent jurisdiction, the validity, legality, and enforceability of the remaining provisions shall not in any way be affected or impaired thereby.

Governing Law: This Agreement, entered into in the County of San Bernardino, shall be construed and enforced in accordance with and be governed by the laws of the United States of America and the State of California without reference to conflict of laws principles. The parties hereby consent to the personal jurisdiction of the courts of this county and waive their rights to change venue.

Entire Agreement: The parties agree that this Agreement constitutes the sole and entire agreement of the parties as to the matter set forth herein and supersedes any previous agreements, understandings, and arrangements between the parties relating hereto.

Appendix D

Installing the data and software

GIS Tutorial 1: Basic Workbook includes a DVD containing maps and data. A free, fully functioning 180-day trial version of ArcGIS 10.1 for Desktop Advanced software can be downloaded at **esri.com/GISTutorial1for10-1**. You will find an authorization number printed on the inside back cover of this book. You will use this number when you are ready to install the software.

If you already have a licensed copy of ArcGIS 10.1 for Desktop Advanced software installed on your computer (or have access to the software through a network), do not install the trial software. Use your licensed software to do the exercises in this book. If you have an older version of ArcGIS software installed on your computer, you must uninstall it before you can install the software that is provided with this book.

.NET Framework 3.5 SP1 must be installed on your computer before you install ArcGIS 10.1 for Desktop software. Some features of ArcGIS 10.1 for Desktop software require Windows Internet Explorer version 8.0. If you do not have Internet Explorer version 8.0 or higher, you must install it before installing the ArcGIS 10.1 for Desktop software.

Installing the exercise data

Follow the steps below to install the exercise data.

1. Put the data DVD into your computer's DVD drive. A splash screen will appear.

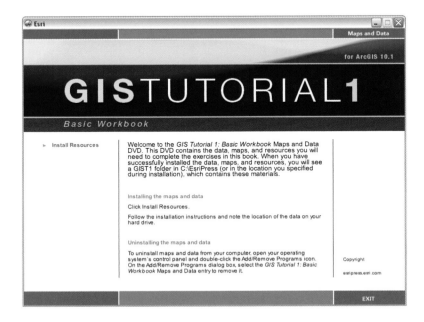

2. Read the welcome, and then click the Install Resources link. This launches the InstallShield Wizard.

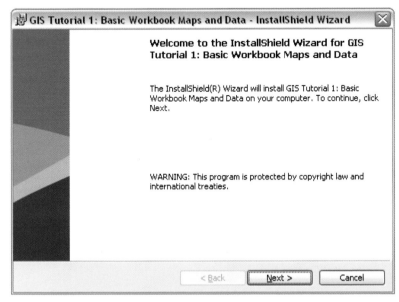

3. Click Next. Read and accept the license agreement terms, and then click Next.

4. Accept the default installation folder or click Browse and navigate to the drive or folder location where you want to install the data.

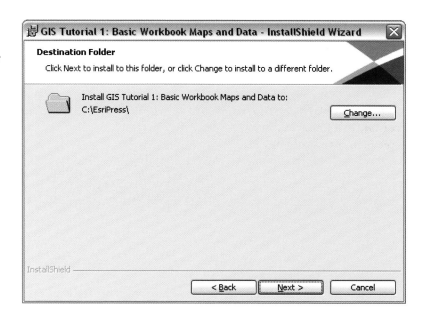

5. Click Next. The installation will take a few moments. When the installation is complete, you will see the following message.

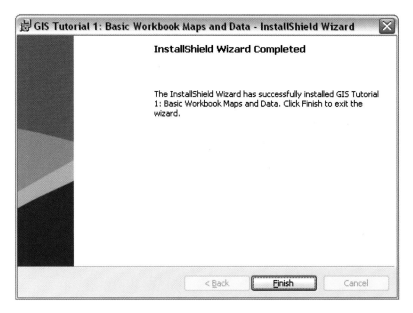

6. Click Finish. The exercise data is installed on your computer in a folder called C:\EsriPress\GIST1.

Uninstalling the data and resources

To uninstall the data and resources from your computer, open your operating system's control panel and double-click the Add/Remove Programs icon. In the Add/Remove Programs dialog box, select the following entry and follow the prompts to remove it:

GIS Tutorial 1: Basic Workbook

Installing the software

Note: If you already have a licensed copy of ArcGIS 10.1 for Desktop Advanced installed on your computer or have access to the software through a network, use it to do the exercises in this book. If you need a copy of ArcGIS 10.1 for Desktop Advanced, you can obtain a free trial version from Esri. This trial version is intended for educational purposes only and will expire 180 days after you install and register the software. The software cannot be reinstalled nor can the time limit be extended. It is recommended that you uninstall this software when it expires.

Follow these steps to obtain the 180-day free trial of ArcGIS 10.1 for Desktop Advanced:

1. Uninstall any previous versions of ArcGIS Desktop that you already have on your computer.

2. Check the system requirements for ArcGIS to make sure your computer has the hardware and software required for the trial: **esri.com/arcgis101sysreq**.

3. Download 7-Zip or WinZip software. You will need to extract the ArcGIS 10.1 for Desktop Advanced files once they have been downloaded to your computer.

4. Install the Microsoft .NET Framework 3.5 Service Pack 1 from the Download Center on the Microsoft website.

5. Go to **esri.com/GISTutorial1for10-1**. Select the option that's right for you and click Next.

Note: You must have an Esri Global Account to receive your free trial software. Click "Create an Account" if you do not have one. When prompted, enter your 12-character authorization number (EVAxxxxxxxxx) printed on the inside back cover of the book, and then click Submit.

6. Install ArcGIS 10.1 for Desktop Advanced.

To download the software

- Extract (unzip) the ArcGIS 10.1 for Desktop Advanced files once they have been downloaded to your computer.

- Go to the location of the extracted files and run ESRI.exe.

To install the software from a DVD

- Insert the DVD into your computer.

- The Quick Start Guide will run automatically. If not, go to My Computer and double-click your DVD drive to run ESRI.exe.

To the right of ArcGIS Desktop, click Setup. Click Next and accept the license agreement. Choose Complete Installation and click Next. Click Next to accept all default settings; then click Finish.

7. Authorize your trial software.

   - Once the ArcGIS Administrator Wizard dialog box opens, select Advanced (ArcInfo) Single Use in the ArcGIS Desktop section, and then click Authorize Now.

   - Continue to click Next until you are prompted to enter your 12-character authorization number (EVAxxxxxxxxx), which you can find printed on the inside back cover of the book. Click Next.

   - Select "I do not want to authorize any extensions at this time." The trial software automatically authorizes the extensions.

   - On the next screen, verify that the extensions are listed on the left under Available Extensions. Click Next.

   - After the authorization process is complete, click Finish.

   - Click OK to close the ArcGIS Administrator dialog box.

8. To start ArcGIS 10.1 for Desktop Advanced, on the taskbar, click the Start button. On the Start menu, click All Programs > ArcGIS and choose ArcMap 10.1.

9. Enable the extensions.

   - Start ArcMap. Go to Customize > Extensions and make sure each extension is checked.

   - Click Close. The software is ready to use.

Assistance, FAQs, and support for your trial software are available on the online resources page at esri.com/trialhelp.

Uninstalling the software

To uninstall the software from your computer, open your operating system's control panel and double-click the Add/Remove Programs icon. In the Add/Remove Programs dialog box, select the following entry and follow the prompts to remove it:

ArcGIS 10.1 for Desktop Advanced